HOW MATH WORKS

HOW MATH WORKS

A Guide to Grade School Arithmetic for Parents and Teachers

G. Arnell Williams

ROWMAN & LITTLEFIELD PUBLISHERS, INC.
Lanham • Boulder • New York • Toronto • Plymouth, UK

Published by Rowman & Littlefield Publishers, Inc.
A wholly owned subsidary of The Rowman & Littlefield Publishing Group, Inc.
4501 Forbes Boulevard, Suite 200, Lanham, Maryland 20706
www.rowman.com

10 Thornbury Road, Plymouth PL6 7PP, United Kingdom

British Library Cataloguing in Publication Information Available

Library of Congress Cataloging-in-Publication Data Available

ISBN 978-1-4422-1874-1 (cloth : alk. paper)
ISBN 978-1-4422-1876-5 (electronic)

♾️™ The paper used in this publication meets the minimum requirements of American National Standard for Information Sciences—Permanence of Paper for Printed Library Materials, ANSI/NISO Z39.48-1992.

Printed in the United States of America

In dedication:

To my friends and colleagues, Joseph Mischel and Debbie Prell, who set marvelous examples for friends and family alike by their bravery, dignity, and positive outlook in the face of life's greatest hardships.

To my California State University, Long Beach Professors Saleem H. Watson and Howard J. Schwartz; both of whom, in freely sharing their insights, gave me a passion for sharing my own.

In celebration:

Of all the men and women (known and unknown) throughout the ages who have contributed to our understanding and use of elementary arithmetic.

Contents

Acknowledgments

THE WRITING OF THIS BOOK has been an exciting journey for me. To attempt to tackle an old and familiar subject with fresh eyes has certainly been a challenge, and throughout I have received assistance from many people on varying levels. First, I would like to express my gratitude to Rachel Black for her patience and support during the earliest years of this project. For their time in plowing through those very early and very rough drafts, I say thanks to Carl Bickford and Callie Vanderbilt. The feedback they provided gave me a better feel for how to reach the book's intended audience.

Those who provided useful commentary during the middle years of this project include Carol Jonas-Morrison, Eric Bateman, Jim Phillips, Judy Palier, Orly Hersh, Andi Penner, Karla Hackman, Russ Whiting, Pete Kinnas, and Michelle Berkey.

The software program Adobe Illustrator proved critical throughout the entire project, and honestly, I'm not sure I could have created the presentation I was trying to build without it. Thanks to Fionna Harrington for suggesting the program and for her technical assistance. Lynn Lane deserves a note of praise in providing much needed expertise in the finer details of Microsoft Word. Thanks to Stephen Gillen for sharing his extensive legal insights.

Barbara Waxer was gracious in offering advice on copyright issues for which I thank her. And Wendy Vicenti deserves a special note of praise for the invaluable feedback she gave on the text from a student's point of view. Bill Hatch was truly generous in lending his artistic talents in the drawing of the Egyptian hieroglyphic numerals.

In any undertaking of this size, an encouraging word or a piece of advice from friends or colleagues proves more useful than they may realize—especially after rejections of book proposals start to occur. For their words of encouragement and advice, I thank Vernon Willie, Angela Grubbs, Sylvia Hensley, Valeria Christea, Kay Brown, Vonda Rabuck, Nancy Mischel, Sherry Nagy, Lynn Jensen, Cheryl Palmer, Nancy Sheehan, Kathleen Chambers, Melanie McAllister, Larry Wilson, Ken Heil, Andrea Ericksen, Greg Ericksen, Laurie Gruel, Shelley Amator, Michelle Meeks, Jon Oberlander, Ann Meyer, Aerial Cross, Laura Kerr, Traci Hales-Vass, Patrick Vass, Susan Grimes, Jeff Wood, and Jenia Walter.

For his willingness to be a constant sounding board as I bounced ideas around as well as for providing insights from his deep well of knowledge, I give particular thanks to Sumant Krishnaswamy.

The last several months have proved the busiest in finishing this work and several individuals provided timely information and advice at key moments. I would like to personally convey my appreciation to Seth Abrahamson and Kate McGowan for giving an involved read to a huge chunk of the text as it neared the final draft stage. Their comments and thoughts proved especially valuable.

A debt of gratitude is also owed those at Rowman & Littlefield who helped me turn my ideas and manuscript into the finished book. I am especially indebted to Patti (Belcher) Davis for believing in me from the start and taking the time to plow through my rough proposal to her as my acquisitions editor—making insightful suggestions on how to improve upon it. If this book proves a good read, it is due in no small part to Patti's advice. A special thanks as well to Suzanne Staszak-Silva for hearing my input and being willing to take up the baton to help me see this project through to completion. Kathryn Knigge was also helpful in going through my final manuscript page by page and figure by figure, for which I thank her.

These acknowledgments would not be complete without me giving a special note of appreciation to my family, in particular my mother, Geneva, and sister, Jennifer, for all they have done for me throughout my life.

And finally, perhaps the greatest thanks of all goes to my students. It has been one of my true joys in life teaching them and it is they who have provided the crucial backdrop on which the core of the ideas in this book were developed and have played out.

G. Arnell Williams
September 2012

Introduction

One of the biggest problems of mathematics is to explain to everyone else what it is all about. The technical trappings of the subject, its symbolism and formality, its baffling terminology, its apparent delight in lengthy calculations: these tend to obscure its real nature. A musician would be horrified if his art were to be summed up as "a lot of tadpoles drawn on a row of lines"; but that's all that the untrained eye can see in a page of sheet music. . . . In the same way, the symbolism of mathematics is merely its coded form, not its substance.

—Ian Stewart, British mathematician and
celebrated popular math and science author[1]

IF YOUR CHILD ASKED why we learn a times table for multiplication but aren't taught one for division, what would you say? It's a basic question. Can you answer it? Are you able to show your child how to do long division, but can't explain why it works? Not just how to perform the method, mind you, but what really makes it go? We all use the symbols {0, 1, 2, . . . , 9} every day: Do you know where they came from or what they are called? What do you call them, and can you explain to someone why we calculate with them instead of with Roman numerals? By the time you finish this book, you will know the answers to these questions and many more, even the most important one that all parents or teachers have been asked: Why is this stuff important?

Put succinctly, this book is for readers who want to know the why in arithmetic—not just the how. If you want to know the context in which arithmetic sits and where the techniques come from, then you have come to the right place. In these pages you will find explained not just how to do multiplication but also what actually makes it tick and how our ancestors tamed it. If you are comfortable in your understanding of the rules of elementary arithmetic, you may still be surprised to learn how much is really involved in making the rules work. If, on the other hand, you are not content in your conceptual understanding of arithmetic and desire to significantly enhance it, then you won't be disappointed.

You may have heard the experts wax eloquent when discussing mathematics, describing it as powerful, mysterious in its reach, even beautiful. Are they serious? To a supermajority of humankind these adjectives are completely invisible when they see mathematics expressed on paper.

My hope with this text is to breathe life into some of that magic and beauty mathematicians rave about when describing their subject. I will attempt to do this by seizing upon them at the fountainhead, for believe it or not, the beauty and power of mathematics are not confined to the higher realms of the subject, but are present in elementary arithmetic right from the start. Conceptual jewels, accessible to you, are available for the taking; and it is my intention to open these up in conversation and view them in the brilliant light of context and history.

While all are welcome to join us on this journey, this book is specifically targeted to address the needs of the general adult reader who, while not being a mathematician or scientist, is nevertheless curious about what mathematics is all about and wants to significantly increase their conceptual understanding of the subject. Hopefully in its reading, you will find that elementary arithmetic is truly spectacular and thereby gain a new appreciation and understanding of the subject in a way that allows you to better deal with the mathematics you might encounter in your life, better explain it to your children, or better understand other math and science books that you may read.

There Is More to Mathematics Than Symbols

A key ingredient in appreciating what mathematics is about is to realize that it is concerned with ideas, understanding, and communication more than it is with any specific brand of symbols. And while symbols form a crucial centerpiece in all of this, they are not the goal in and of themselves.

In terms of using ideas in extremely powerful ways, mathematics holds an exceptional, almost hallowed place. It is no stretch of the English language to say that ideas and reasoning cast in mathematical form are truly something else. The great Galileo is said to have declared that, "Man's understanding where mathematics can be brought to bear, rises to the level even of god's."[2]

It is almost as if ideas set in mathematical form melt and become liquid and just as rivers can, from the most humble beginnings, flow for thousands of miles, through the most varied topography bringing nourishment and life with them wherever they go, so too can ideas cast in mathematical form flow far from their original sources, along well-defined paths, electrifying and dramatically affecting much of what they touch.

For us to dial into this transportability, however, requires that we use symbols—a lot of them in fact (think of the symbols as part of the fluid and the rules of mathematics as part of the riverbed). It is through the use of symbols that human beings can leverage the almost magical ability of mathematics to systematically and reproducibly transform ideas into other ideas, and the need for them appears quickly when we try to answer questions involving quantity.

Why Symbols Are Needed in Arithmetic

People have always had the need and desire to compare and analyze the sizes of collections. How much stuff do we have? How many people are in our settlement? How large is our enemy? Collections, such as these, vary in size and when we get to the point of describing or cataloging these variations in-depth, we are inevitably led to symbolic descriptions. How do symbols help us? Let's take a peek.

Consider a scenario involving two cattle ranchers, each with a large herd numbering into the thousands, wanting to know who has more cows. For the time being, let's assume that no system of numeration has been developed and that they must figure out a way to do the comparison from scratch. How will they be able to prove, beyond dispute, who has the larger herd?

There are several ways to proceed. One involves the ranchers creating a pair of lanes (one for each herd) and then having their ranch hands round up the cows and march them singly down each of their respective lanes in a matching off process. If the herds are of unequal size, one of the ranchers will eventually run out of cows in the pairing. The one with the excess of cows can then conclude that he has the larger herd. While this method certainly works in determining who has the larger herd, it could be very difficult to accomplish in practice. There are better ways.

Another method involves using two carts (one for each herd) and a large collection of small rocks. Each rancher's herd is now measured by going out into their respective pastures and placing a rock in their respective carts for each cow. Once each herd has been measured in this fashion, it is a much simpler matter to bring the carts in close proximity and pair off the small inanimate rocks than it is to round up and pair off two sizeable herds of huge, living, smelly animals. The ranchers can obtain the same information as with the first method but this time in a much more convenient manner.

Each rock in the collection has acquired a new meaning—rather than simply being a rock, it now stands for a cow. Or put another way, each rock has become a symbol.

Two great strides are gained by taking this simple step. First, it is clearly much easier and more convenient to match off small inanimate rocks than it is matching off hundreds of large animate cows, each with its own agenda. Second, using the rocks as symbols has opened up a vastly superior way of comparing collections. Given that existence of an object is what counts in whether a rock is placed into the cart, there is nothing that prevents the ranchers from comparing other things that exist besides two herds of cattle. They could just as easily use these carts and rocks to compare the sizes of two groups of people, two neighborhoods of houses, two forests of tall trees, and so on. For many of these situations, the two lanes method is impossible to use at all. Large houses or tall trees cannot be easily rounded up, marched down lanes, and paired off. So we see that the method with rocks is not only more handy than the method with lanes, it also gives the ability to compare a greater variety of objects.

Since they are in the mood, can they find any symbols more convenient than using rocks? Absolutely! If the ranchers had some sort of portable writing system, they could replace the rocks in the carts with written tally marks. For instance, they could use any of the following sets of marks: |, X, O, or +. If they chose to use |, three rocks in a cart would be represented as: | | | .

Once each had done his separate tally of his respective group, the ranchers could simply compare or match off the written symbols and no longer be burdened by pulling heavy carts full of rocks. And since tally marks can be created at will whereas rocks cannot, tally marks can, in theory, measure much larger collections without as great a concern for supply issues.

Each of these improvements can be looked upon as a "technological" breakthrough in how collections are measured, and it is clear to see that the method of indirect comparison, in this case using symbols that stand for the objects being counted, has decisive advantages over directly using the objects themselves. Throughout this book, we will see that in mathematics symbols are absolutely necessary.

Symbols Are Important in Language as Well

The need for using symbols is not unique to mathematics. Other systems critically depend on them as well. The most familiar of these are spoken languages. Spoken languages are systems that use sounds as symbols. They give us the remarkable ability to describe and communicate with easy to produce sounds as opposed to trying to do so by reconstructing, out of thin air, the physical objects, events, and ideas that we wish to describe.

In other words, spoken languages give us the ability to represent a substantial portion of life through the use of nothing but sounds.[3] Speaking allows us to take our inner thoughts and share them with others by simply making sounds with our vocal chords. A song consisting of nothing but sounds can bring people to tears or motivate them to action. Think of the organized sounds of speech serving as part of the "fluid" for transporting thoughts and emotions just as numerical symbols are part of the "fluid" for communicating quantitative information in mathematics.

Cognitive scientist George Armitage Miller states, "The evolution of language enabled many individuals to think together. Social units could form and work together in novel ways, cooperating as if they were a single super ordinary individual."[4] One emergency worker can, for example, using mostly sounds, organize ten men to lift a heavy car off a victim; acting together, if you will, as one super human. Spoken languages give dramatic demonstration to the fact that using symbols to represent ideas can be extraordinarily powerful.

Negative Side Effects of Using Symbols

Despite being essential for the expression of mathematics, symbols are notoriously bad in that they can quite naturally mask what is happening conceptually. This unfortunate side effect of using symbols is one of the central issues that math education must overcome.

Since symbols in mathematics rarely look like what they describe (e.g., the tally marks or rocks discussed earlier look nothing like the cows they represent), using them necessitates a temporary separation between the problems and motivations and the method of solution. This in itself isn't a problem. The problem arises when this separation is taught as being the natural state of affairs—or worse yet, the only state of affairs.

When this happens, it can become difficult for students to acquire a proper perspective of what the symbols are really doing for them, and since students are people and not machines, this has psychological implications that can prove fatal to their understanding, and forever affect their attitude toward the entire subject of mathematics.

On the other hand, for the symbols to be most effective, they must be allowed this separation (i.e., the rocks and tally marks must be allowed to represent other things besides the cows). Only when the symbols are allowed to free themselves from their origins can they really fly and open up whole new worlds to those who use them.

So we have an interesting paradox.

The very strength of mathematics, its use of efficient and unfettered symbols and procedures to express and transform ideas, is also its greatest pitfall in terms of conceptual and contextual understanding. This paradox is real and unavoidable. The situation can be likened to the difficulties with faithfully capturing in writing what someone said or thought.

Issues with Making Communication Visible

Imagine this entire chapter without any punctuation marks, spaces between words, or capitalization. The quote by Ian Stewart at the beginning of this chapter would read as:

oneofthebiggestproblemsofmathematicsistoexplaintoeveryoneelsewhatitisal laboutthetechnicaltrappingsofthesubjectitssymbolismandformalityitsbaffling terminologyitsapparentdelightinlengthycalculationsthesetendtoobscureitsreal natureamusicianwouldbehorrifiedifhisartweretobesummedupasalotoftadpoles drawnonarowoflinesbutthatsallthattheuntrainedeyecanseeinapageofsheetmusic inthesamewaythesymbolismofmathematicsismerelyitscodedformnotitssubstan ceianstewart

as opposed to

One of the biggest problems of mathematics is to explain to everyone else what it is all about. The technical trappings of the subject, its symbolism and formality, its baffling terminology, its apparent delight in lengthy calculations: these tend to obscure its real nature. A musician would be horrified if his art were to be summed up as "a lot of tadpoles drawn on a row of lines"; but that's all that the untrained eye can see in a page of sheet music. . . . In the same way, the symbolism of mathematics is merely its coded form, not its substance.—Ian Stewart

It is easy to see that punctuation marks, spaces between words, and capitalization are a tremendous help in making passages such as Stewart's quote more convenient to read and understand. These are issues of writing not speech; when statements are spoken they come at us in a very different fashion than they do when we attempt to make them visible.[5] A major task in writing then is to design it in such a way that it recaptures, as much as possible, what is being spoken or thought.

If a statement is spoken, the speaker adds tone, facial expressions, eye contact, and gestures to the statement to convey meaning. If the statement is a silent thought, then the thinker has context, images, and emotions in mind when he is thinking the thought.

In either case, when the statement is penned to paper, these extras (tone, gestures, mental images, emotions, etc.), the nonverbal cues, if you will, are lost. The writer can make an attempt to recapture these; by adding context, formatting, careful phrasing, and through sprinkling the written text with punctuation (we will call these the "writing extras").[6]

If the statement is a question, then the question mark symbol can be used to communicate this. If the speaker or thinker wanted to express the statement with emphasis, then the written form can be given an exclamation point, quotation marks, boldface or italics, bullets, set in a tabular format, and so on.

The writing extras help to clarify the speaker or thinker's intent—making them a contextual and conceptual illumination of sorts. These extras rarely if ever do this perfectly but serve as a tremendous aid in helping written communication stand on its own.

Conceptual and Contextual Illumination

We might ask a similar question of explaining mathematics: How do we present mathematical material such that the average reader, with the appropriate background, can understand in a substantial way what is truly being communicated about the mathematics, and then have that reader also gain a true appreciation of the subject from that understanding?

It is important to recognize that the written expressions of mathematics have many of the same issues that language writing has in trying to effectively convey context and meaning. Moreover, many of these issues are of a far greater intensity in mathematics due to the simple fact that unlike language writing, which is intimately connected with the spoken language of the reader, mathematical writing seems more like a foreign language.

This is an extremely serious problem in mathematics education as the celebrated author Lancelot Hogben has alleged, with historical justification: When a subject loses contact with the common man it runs the risk of becoming a superstition.[7] And it is a problem that really exists to some degree at all levels of the subject. Even exceedingly competent mathematicians and scientists face this problem to some extent when trying to understand and utilize an area of mathematics with which they are not familiar.

If it is to make the wonders and beauty of mathematics more generally accessible to the public, mathematical writing, far more so than language writing, needs to be significantly enhanced through conceptual and contextual illumination. Doing this properly is an enormous undertaking and can be likened in scope, perhaps, to the difficulties in making many of the scenic marvels in the western United States, at one time reachable only by intrepid explorers, accessible to the average citizen. Accomplishing this necessitated massive construction projects involving the creation of thousands of miles of roads, hundreds of bridges, whole cities or towns, facilities, signage, and an infrastructure involving thousands of people to manage it all.

It is with an eye toward contributing, in some small part, to this massive undertaking that *How Math Works* was written. Being an experiment in exposition, it aims to provide in a deliberate fashion some of that critically needed conceptual and contextual illumination.

Final Words

The universe is very generous in that it allows us to gain remarkable advantages by converting or substituting one set of things into or for another set of things. A magnificent application of this is the depiction of meaningful ideas/objects whether physical or abstract in the form of coded symbols (i.e., symbols that look nothing like what they describe). It is such a natural thing to do in mathematics that symbols have indeed become the public face of the subject. The great advantage to employing them is that it is far easier and more advantageous in general to use and manipulate symbols than it is to use and manipulate the things they stand for. This advantage gives human beings the breathtaking ability to accomplish absolutely astounding feats from extremely comfortable positions.

Mathematics, however, is more than just the manipulation of symbols, and the goal here is to shed critical light on this fact. Our discussion will revolve around what have historically been considered the five fundamental operations of elementary arithmetic. These include the standard four of addition, subtraction, multiplication, and division, as well as that of the representation of quantity, often referred to as numeration. We will focus on whole numbers exclusively.

This material may seem too shallow to base an entire book around but, as you will discover, nothing could be further from the truth. There is great depth, beauty, and genius inherent in these five operations and the framework surrounding them, and we will attempt to bring awareness to this fact by flooding the conversation with conceptual and contextual illumination.

Many of the questions we will address include:

- What does it mean that our number system is "base ten"? What is the significance of the notions ten, hundred, thousand, . . . , million? (See chapter 1.)

- What do language writing and mathematical writing have in common? Can any of this be useful in illuminating issues in mathematics? (See chapter 2.)
- Where does our number system come from? (See chapter 3.)
- What is the modern significance in math education of the ancient tool called an abacus? (See chapters 3, 5, 6, and 8.)
- Why do some number systems, such as ours, have a zero and other number systems, such as Roman numerals, have none? What are our numerals called? (See chapter 3.)
- Why do we care about mathematics? (See chapter 4.)
- Where do the vertical numeral formations we use in addition, subtraction, multiplication, and division originate? (See chapter 5.)
- What is the true significance of a times table? Are there tables for the other operations? (See chapters 5, 6, 7 and 8.)
- What makes the multiplication algorithm really tick? (See chapter 7.)
- What are three major interpretations we can give to division? Why is division by zero undefined? (See chapter 8.)
- What is going on with the long division algorithm? (See chapter 9.)
- Why did the numerals we use today replace all of the others? (See chapter 10.)
- What was arithmetic education like 500 years ago? How did medieval Italian merchants and eighteenth-century Swiss educators influence elementary arithmetic education in America? (See chapter 11.)
- How do numbers help illuminate our world? What do measurement and counting have in common? (See chapter 12.)

It is now time to begin our celebration of this most fundamental of subjects. I am excited! Hopefully you are too and will enjoy the journey.

I

THE RELEVANCE OF THE PAST

History is the most fundamental science, for there is no human knowledge which cannot lose its scientific character when men forget the conditions under which it originated, the questions which it answered, and the function it was created to serve.

—Benjamin Farrington, Irish writer,
classicist, author of *Greek Science*[1]

I am sure that no subject loses more than mathematics by any attempt to dissociate it from its history.

—James Whitbread Lee Glaisher, English mathematician,
astronomer, editor of the *Messenger of Mathematics*
(inspiration for the American Mathematical Society)[2]

1

Tools of the Intellect

Neither the bare hand nor the unaided intellect has much power; the work is done by tools and assistance, and the intellect needs them as much as the hand. As the hand's tools either prompt or guide its motions, so the mind's tools either prompt or warn the intellect.

—Francis Bacon, English statesman, lawyer,
philosopher and voice of the Scientific Revolution[1]

IN NOVEMBER 1890, the noted English psychologist James Ward boldly stated of mathematics education that: "The individual should grow his own mathematics just as the race has had to do. But I do not propose that he should grow it as if the race had not grown it too. When, however, we set before him mathematics in its latest and most generalized, and most compacted form, we are trying to manufacture a mathematician, not to grow one."[2]

An interesting idea indeed, but it was already very old even by Ward's time. Believe it or not, traces of this idea can be found way back in the 1600s in the writings of the great Czech educator John Amos Comenius, as well as many others since (and a few before). In this chapter, we aim to implement a portion of this grand vision of math education by attempting to grow the reader's understanding of basic numeration in a natural and contextual manner as opposed to manufacturing it artificially. Let's see if it works.

We begin with a simple question: What exactly is a numeral? Is it the same thing as a number? If not, what's the difference? Put to mathematics, this may seem like a subtle question but it is essentially a question about the difference between a thing and the symbolic representative of that thing. For example, it is easy to see that an actual physical house is vastly different from the five letter word "house" that represents it in English. The word in fact doesn't look anything like the structures it describes.

In a similar fashion, numerals are symbols that we use to represent numbers but aren't the actual numbers themselves. Defining exactly what a number is can be tricky. The good news is that our intuitive notion of what numbers are is more than adequate for our purposes here. For example, whether or not we can give an exact definition of the number three doesn't prevent us from recognizing when a collection has three objects. In this text, we will often blur the distinction between number and numeral, but this should cause no difficulties.

It is worthwhile noting, however, that when the distinction between the two has been blurred this has sometimes led to important mathematical discoveries (we will see this happen in chapter 3 when we illuminate the discovery of zero), while in other cases it has led to untold conceptual difficulties and superstitions (some of which also came with the discovery of zero).

In any event, it is useful to keep in mind that there is the concept of a number (say three) and many different ways to represent that number (three, tres, | | |, 3, etc.); just as there are many different ways to represent the concept of greeting someone (e.g., hello, hi, buenos dias, ni hao, konnichiwa, guten tag, namaste, and so on).

Organizing Tally Marks

Throughout this book, we will be primarily concerned with numerals that represent a number by some sort of written mark. When we decide to represent numbers this way, we gain advantages in efficiency—over other methods such as direct comparison or using rocks. Moreover, by using visible marks we acquire the ability to more easily lay out side by side the symbols for the various sizes; which allows us to analyze how different magnitudes compare in a visual way. Implementing this scheme by going from smaller values to larger values yields the following system of tally mark arrangements:

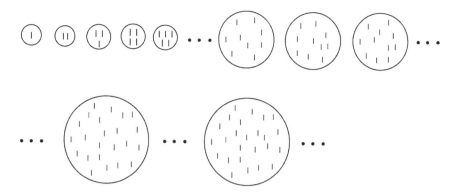

Now, let's say we wanted to count a collection of ten apples. We could do this by simply jotting down a tally mark for each apple and this might look like: | | | | | | | | | |. Since the number of tally marks is the determining factor in conveying size, not their horizontal or vertical appearance, this arrangement is equivalent to the one in the previous diagram given by: ⊙. We see then that the number ten is represented in the system of tally mark arrangements listed above.

While giving us greater leeway than the original methods of comparing size (via lanes and rocks), using tally marks still has some serious inconveniences. This can be seen when tally marks are used to represent relatively large collections, say for instance one thousand.

Think about it, placed side by side would you be able to quickly distinguish the difference between a collection containing one thousand tally marks and a collection that had one thousand and three tally marks? You would be a rare specimen indeed if you could. Humans simply aren't built to quickly gauge such differences on sight. We know that we can resort to a matching off process but this is laborious and unpractical. Can we build tools of some sort to help make such differences more readable to us on sight?

Societies throughout the ages have faced this question and chose a variety of ways to deal with it. A natural approach, when the number of tally marks became large enough, involved rearranging them into equally sized groups. If we apply this approach to the largest two collections shown in the bottom row of the tally mark arrangements, we obtain many possibilities. Three rearrangements are listed here:

Groups of five:

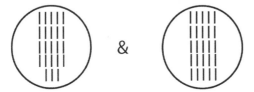

Groups of seven:

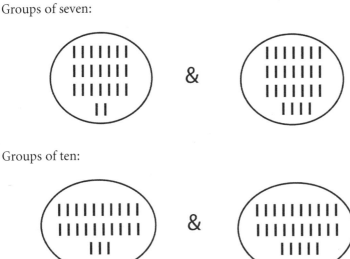

Groups of ten:

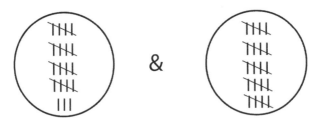

Each of these rearrangements allows us to determine, more quickly, which collection of the two is larger (twenty-three versus twenty-five). Such tallies are frequently done in counting votes and the like; and to simplify matters even further, each of the rows of equal size is often struck out by a diagonal line, which is also included as a tally mark in the count. Thus our rearrangement into groups of five can be written as:

Groups of five (rows slashed):

Similarly, our seven and ten groupings can be written as:
Groups of seven (rows slashed):

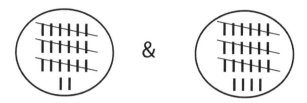

Groups of ten (rows slashed):

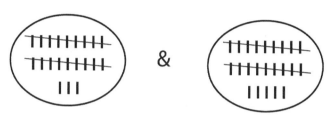

More conveniently, we could replace each of the rows of constant size by a new symbol. Thus, for example, in the first case we may replace the rows of size | | | | | by the symbol (F). Doing so compresses the symbols used to:
Five coins:

Similarly, replacing the groups of sizes | | | | | | | and | | | | | | | | | | by the respective symbols (S) and (T) in the latter two examples yield:
Seven coins:

Ten coins:

Compare the readability of the coin diagrams with the original ungrouped tally mark diagrams. The size of the groupings chosen in this manner is called the base of that system. Thus the reorganizations into groups of five, seven, and ten correspond to systems of base five, base seven, and base ten, respectively.

While there is complete freedom in which system of grouping or base to use, most societies in the past chose to group in fives, tens, or twenties as opposed to other groupings such as sevens or twelves. Why is that?

The common belief is that these choices were made due to the nature of human anatomy. The human beings devising these systems had five fingers on one hand, ten fingers on two, and ten toes on two feet. It seems reasonable that these appendages would be used to aid in the counting, and that people would naturally choose groupings that matched the number of fingers and/or toes in a one to one fashion.

A good test of this theory could occur sometime in the future: If we were to ever meet an alien species of beings with say seven fingers on each hand and seven toes on each foot. The theory would predict that popular bases for this species would be seven, fourteen, and twenty-eight (assuming we didn't destroy each other first before we could find out).

We will use base-ten grouping throughout the book. Accordingly, any collection of tally marks numbering less than ten is left as it occurs, thus we will leave the arrangements { | | | | | , | | | | | | | | } as they are but will replace the arrangement with ten tally marks { | | | | | | | | | | } by the (T) symbol or the arrangement with twelve tally marks { | | | | | | | | | | | | } by the coins (T) | | (where we have replaced ten of the tally marks by the ten coin symbol, leaving as is the two left over). Whenever a group of ten tally marks is encountered in any configuration we will so replace them by the (T) symbol.

Using these grouping rules now gives us the ability to quickly and easily distinguish between a coin configuration containing, say, eighty-three tally marks and one containing eighty-five. Not the case, however, if we leave these coin configurations ungrouped, as eighty-three |s and eighty-five |s, respectively.

Using the rules for grouping in tens: eighty three and eighty five tally marks are represented, respectively, by the following more easy to read coin arrangements

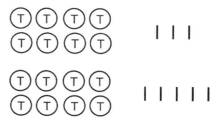

This problem of comparison is not unique to the |s. The essence of the difficulty lies with the proliferation of symbols. It is hard for us to quickly compare eighty-three and eighty-five of any type of symbol. So if we have eighty-three and eighty-five (T)s, we will still have the same problem.

The process of gauging or measuring the size of a collection on sight without counting is called subitization. Think about it, you can probably recognize at most four or five objects at a glance. Ask yourself, and be honest now, can you look at a collection of thirteen objects and recognize instantaneously that there are exactly thirteen of them without counting or rearranging them in some way?

Some experimental results suggest that our ability to recognize a certain number of objects on sight is an innate talent that we possess from a young age. In a particularly interesting study reported in *Nature* magazine in 1992, reaction times of four- to five-month-old infants were investigated by a University of Arizona researcher. The infants paid much longer attention to situations involving size that violated what they evidently expected to see. For instance, a puppet was shown to the child and then hidden behind a screen. Then a second puppet, in full view of the infant, was placed behind the screen. When the screen was removed and only one puppet was present, as opposed to the two puppets, the infants paid much longer attention to this situation than they did to the expected situation of two puppets being present. Similar observations were observed, for instance, when two puppets were expected and three were shown. This suggests that the infants could distinguish between one and two and between two and three (and that they even understood that *one* plus *one* should be equal to *two* not one or three).[3] Other studies have since both confirmed and questioned these findings.[4]

We have made the arrangements represented by:

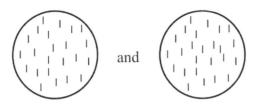

and

more readable by invoking our base-ten grouping rules and rewriting them as:

Can we do the same if we are confronted with twenty-three (T)s and twenty-five (T)s as shown here?

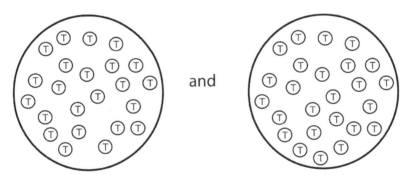

and

An obvious way to make this situation more readable would be to similarly represent a group of ten of the (T) coins by a new symbol. This we do and get:

(H)   which replaces

The "H" of course stands for one hundred and using this new grouping allows us to rewrite the twenty-three tens and twenty-five tens, respectively, as:

The symbols on the right represent the larger number of two-hundred and fifty versus the smaller value on the left of two-hundred and thirty. This form of writing the values is easy to read and once again allows us to immediately recognize which number is greater.

If the sizes are great enough, the problems we have with the (T)s will reappear again with the (H)s unless we also group them and so on. Thus continued grouping into higher coin denominations is an obvious choice if we want to handle larger and larger numbers more conveniently. Historically, as societies grew more complex they had to deal with ever-increasing numbers, leading to the creation of not just one type of coin numeral but to many types—or to a system of coin numerals. The natural progression of our system is listed here:[5]

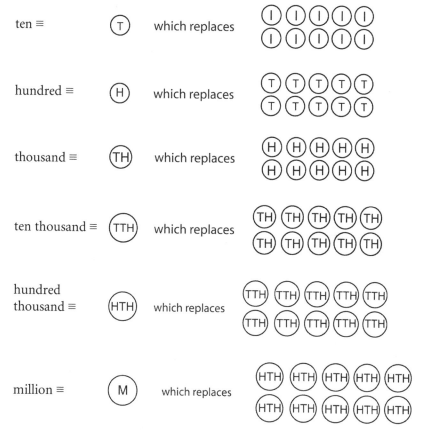

Imagine trying to understand, without any grouping, the individual tally marks represented by:

This number in English reads as two million, two thousand, one hundred twenty-one and in grouped form we can grasp what value it represents with ease. Without the aid of grouping, we are as helpless mentally to grasp its value as a logger would be in a forest without the aid of his chain saw or ax.

Representing numbers by grouping doesn't save us the trouble of having to do the initial count, however. If there are eight hundred thirty-five objects randomly assorted, the first person to learn this must perform the actual tally. However, once the result is obtained, grouping allows her to rearrange the tally marks, via the methods discussed here, and record them in a more accessible fashion for future use. That is, this grouping saves others, including herself, who need to reference the information, from having to perform the long count all over again. If the eight hundred thirty-five objects are arranged in a neat pattern, we may be able to determine the number of objects without actually counting each one; we'll talk more about how to do that in chapter 6.

This simple process of rearranging larger groups of tally marks into smaller sets of coins hints at the fantastic capabilities for transformation that ideas about quantity can acquire when we represent them by certain types of visible marks. We can systematically with ease take the visible marks places that it would be very difficult, if not impossible, to take the things that they stand for. This allows us to literally reshape ideas in reproducible ways (one thousand individual tally marks being instantaneously morphed into the single symbol (TH) while the thousand individual physical items that the symbols stand for remain unchanged)—refashioning them so as to gain huge conceptual and, as we shall soon see, computational advantages while not losing any essential content.

It's just like money. Money allows us to equate all sorts of different things and represent them by a single dollar value. Eight hundred dollars can represent a rent payment, the price of a television set, the price of a computer, the wages for forty hours of hard work, a car repair bill, or the price of a meal at a campaign fund-raising event. These different things become equivalent in the sense of monetary value. This equivalence allows a laborer to systematically transform his forty hours of hard work into a television set, a rent payment, a set of new brakes, or a computer. Money, in a sense, acts like a liquid by allowing his forty hours of sweat to seamlessly flow into another form to great advantage (in this case a form involving the necessities or luxuries of life). In fact, assets in the form of cash are often called liquid assets. Symbols in arithmetic act like a currency for quantitative ideas.

Ancient Numeral Systems

We have discussed one possible way of making the sequence of sizes for collections more readable. There are a myriad of others. One of these is the Egyptian hieroglyphic system. The hieroglyphic method of numeration is similar in spirit to the coin system of numeration discussed earlier. The following table shows the coin denominations and their hieroglyphic equivalents.

Coin Symbol	Hieroglyphic Symbol
ⓘ one coin	| hierglph one
Ⓣ ten coin	∩ hierglph ten
Ⓗ hundred coin	◎ hierglph hundred
(TH) thousand coin	hierglph thousand
(TTH) ten thousand coin	hierglph ten thousand

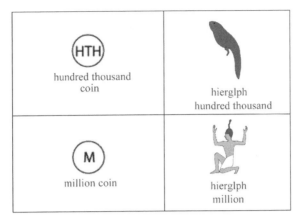

hundred thousand coin	hierglph hundred thousand
million coin	hierglph million

Permission to use hieroglyphic symbols granted by William Hatch.

Using this system, the number two million, two thousand, one hundred twenty-one would read as:

The number twenty-three thousand, three hundred twenty-two in hieroglyphics would read as:

Another ancient system of numeration is the well-known Roman numeral scheme. The Roman system uses "*I*" as its basic tally symbol. Thus the side-by-side layout of magnitudes with this symbol is given by:

I II III IIII IIIII . . . IIIIIIII IIIIIIIII IIIIIIIIII . . .

It is clear that the Romans faced the same problem with the proliferation of symbols as we did earlier. The Romans chose to also deal with this problem by grouping—converting a certain collection of tally marks into a new symbol; but they did so in a manner somewhat different from our coin system.

Their first grouping occurred at five, with the symbols *IIIII* being replaced by the symbol *V*. However, instead of continuing to group in fives and creating a new grouping symbol for *VVVVV*, the Romans created a new grouping symbol for only two of them, replacing the symbols *VV* by the symbol *X*. This alternation between five grouping and two grouping continues in the Roman scheme; indicating the use of fingers and hands (which come in fives and twos, respectively).

The following is a list of Roman symbols and how they represent magnitudes up to one thousand:

(One): *I*
(Five=Five ones): *IIIII* ≡ *V*
(Ten=Two fives): *VV* ≡ *X*
(Fifty=Five tens): *XXXXX* ≡ *L*
(Hundred=Two fifties): *LL* ≡ *C*
(Five Hundred=Five hundreds): *CCCCC* ≡ *D*
(Thousand=Two five hundreds): *DD* ≡ *M*

Unlike our coin scheme which is purely additive (i.e., we add the coin values in a configuration to obtain the value represented), Roman numerals can also employ a subtractive feature. Thus for instance one may write:

four as *IIII* (four ones) or as *IV* (five minus one)
nineteen as *XVIIII* (fifteen plus four) or as *XIX* (ten plus ten minus one)
nineteen hundred as *MDCCCC* (fifteen hundred plus four hundred) or as *MCM* (one thousand plus one thousand minus one hundred)

The convention in general is that larger value symbols are written to the left of smaller value symbols. If a smaller value symbol is to the left of a larger value symbol, it means subtraction of the smaller value from the larger. Thus *XXXI* means thirty-one whereas *XXIX* means twenty-nine. Numbers larger than one thousand can be dealt with by placing a bar over the number, which means to multiply by one thousand. Thus $\overline{V}$ means five thousand and $\overline{L}$ means fifty thousand.

It is worth noting, however, that much variation has existed in some of these rules throughout the centuries. For instance, the subtractive principle appears to have been rarely practiced in ancient Rome itself, and several methods, in addition to the one discussed above, have been used to represent numbers greater than a thousand.[6]

Roman numerals, of course, still find considerable usage today. They serve as visually elegant representations of numbers, and are sometimes used in movie credits, on clocks, for the labeling of Super Bowls, and as page numbers in the prefaces and introductions to books. They are not used, however, to perform calculations.

Six hundred seventy-two is represented in the three systems as follows:

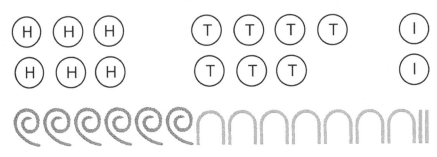

DCLXXII

What is the base of the Roman numeral system? Since two types of groupings are used, which has the priority? Is it a base-two system, a base-five system, or something else? Essentially, the Roman numeral system is what might be called a modified base-ten system. That is, all the base-ten groupings of our coin system (ones, tens, hundreds, thousands, ten thousands, etc.) are present in the Roman system (using bars for values larger than a thousand) but there are extras: five, fifty, five hundred, and so on.

As a standard of comparison, think about the American monetary system which is a decimal or base-ten system with modifications—that is, we have the base-ten denominations of the penny (the equivalent to the basic tally mark), the dime, the dollar bill, the ten dollar bill, the hundred dollar bill, and so on, but we also have the nickel, quarter, five dollar bill, and twenty dollar bill.

These extra or interior denominations help keep the change in our wallets and purses more manageable and so too do they help keep the strings of Roman numerals down to a more manageable size. This is an advantage of the Roman system over the other two we have discussed. This can be clearly seen in the representation of six hundred seventy-two in the three systems (where the coin and Egyptian systems require fifteen total symbols while only seven Roman numerals are needed).

Another ancient system of numeration, which we will briefly visit later in chapter 10, worked by grouping in packets of sixty. That is, instead of being base ten, the system was what we might call a modified base-sixty system. This is the famous sexagesimal system of the Sumerians, and subsequently the Babylonians, developed more than four thousand years ago. The system was a mix of additive grouping and place-values, thus differing in a fundamental way from the three systems discussed earlier.

Awkward as it may sound, this system was so versatile, due to the place-value component, that remnants of it survive to this very day. For example, in our methods of reckoning time, we group both minutes (the small or "minute" part of an hour) and seconds (the "second" small or minute part of an hour) in packets of sixty.

Conclusion

We have progressed far from the pedestrian task of comparing two herds of cattle. The work done on the ranch clearly has had far-reaching implications. Starting from the methods developed there, we now have a coin numeral system allowing us to conveniently represent the quantity in any discrete collection we are likely to ever encounter in daily life. Moreover, we can use these representations to compare the sizes of various collections and communicate these results in a visible way. The ranchers in satisfying their curiosity to know which of them had the larger herd were actually touching upon universal situations that can all be dealt with in a similar fashion—meaning that they were, in a sense, able to dial into eternity from down on the ranch.

The requirements for using a manageable set of symbols are based in large part on psychological needs. We simply cannot easily distinguish the differences between collections of even modest size unless we arrange them in some fashion. Ultimately, however, we are still fighting a losing battle. Given that numbers go on without end, even our technique of grouping will result in a proliferation of symbols that will eventually overwhelm, as we try to describe larger and larger numbers.

The coin system or even the place-value system we use today, while simple to learn and rich enough to describe most of the numbers we need in everyday life, are not so well suited to describe some of the numbers that scientists and engineers need on a daily basis. Consequently, they often use other notational schemes to keep the number of symbols down to a manageable size. Scientists often use a system called scientific notation while others often use a slightly different version called engineering notation.

Finally, the ability to share information in a visible way turns out to be highly nontrivial. In language, the ability to communicate visibly turned out to be revolutionary; perhaps as revolutionary a thing as there has ever been designed by the hand of man. In the next chapter we discuss some of the gains that we acquire from communicating this way.

2

The World in Symbols

This marvelous invention of composing out of twenty-five or thirty sounds that infinite variety of expressions which, whilst having in themselves no like-ness to what is in our mind, allow us to disclose to others its whole secret, and . . . all that we imagine, and all the various stirrings of our soul.

—Port Royal Grammarians (1660), French grammarians, proponents of a universal grammar[1]

Writing not only helps us remember what was thought and said but also in-vites us to see what was thought and said in a new way.

—David R. Olson, contemporary Canadian linguist and author of *The World on Paper* (1994)[2]

BEFORE CONTINUING WITH OUR tour through elementary arithmetic, we pause briefly to turn our attention to the symbolic systems that we use to com-municate with each other every day. It is well known that human languages are of monumental importance to most of what we do, and despite the fact that we are intimately familiar with how to use them (our own at least), much about them remains a mystery. How do languages develop?[3] How do children truly come to learn languages?[4]

Whatever the answers to these questions, we know that both language and mathematics are used by human beings, in part, to help them better under-stand and better steer their way through a complex world. Is there any com-mon ground between the two—common ground that might prove useful to a better understanding of both? Surely there must be a lot. Mining that common ground for a few grains of conceptual insight is the focus of this chapter.

Everyday communication comes in two main flavors: spoken and written. Each form represents an important category by which we are able to symbolically represent certain aspects of the world. As this book unfolds, these categories will prove useful in conceptually understanding important features regarding numeration and calculation in elementary arithmetic.

Speech is the more natural of the two and by itself already shows a couple of important features of symbolic systems in general:

1. *They allow us to represent one set of objects by another more convenient set of objects.* In the case of speech, representing complicated and hard to reproduce human activity more conveniently through the use of easy-to-produce sounds.
2. *They can come in a wide variety of forms.* Several thousand different spoken languages exist in the world today. They differ in the details, yet they all exist for the same basic purpose: to facilitate communication using sound signals.

Yet as powerful as speech is, it is missing two important ingredients sometimes needed in communication: permanence and static visibility. Sounds are temporary and invisible, and they quickly disappear. All of that coded meaning in those sounds vanishes almost immediately after they are created (the problems with lack of permanence and static visibility are not limited to speech—for as we saw earlier, the human need to build systems to determine and permanently record the varying sizes of their possessions eventually leads to the creation of numerals [e.g., coin numerals in our case]).[5]

How did people address these drawbacks of speech? One is naturally led to visible marks. The gain in permanence using visible marks as compared to using sounds is astronomical; for example, the words in a tragedy written by William Shakespeare in the early seventeenth century remain more than 400 years later, but the sounds made by human beings alive during his time vanished seconds after they were made. A way to overcome these limitations of speech is to figure out some way to represent communication by using these more permanent symbols. People did this, and their efforts to incorporate visible marks into communication can be roughly divided into two broad categories:

1. *Pictorial Based Systems*: Systems in which the visible marks represent ideas through pictures (that often look like what they describe) and have no direct relation to the spoken language of the users.
2. *Language Based Systems*: Systems in which the visible marks represent some component of the spoken language of the users.

The first type of system is much older. Humans have been sharing information pictorially for tens of thousands of years.[6] We still do so today. Examples include using drawings rather than words to represent men's and women's restrooms, drawings of a deer to represent deer in the area, or using illustrations in road signs to indicate the presence of children and to drive with caution. Symbols of this type are called pictograms.

In general, as in these cases, the symbols are used to communicate through pictures rather than through the words of a language. Being independent of speech, nearly everyone, regardless of native tongue, can quickly understand what the symbols mean.

Photograph and design by author.

This sign is a pictogram. A person does not have to understand English to understand what it means. Using pictures not words, it communicates the following information: *Do not park in this location; if you do your car will be towed and it will cost you money to get it back.*

It would appear to be impossible to capture the full range of human experience using only pictorial-like systems (although this has not been proven).[7] Imagine the difficulty in trying to accurately describe to someone everything that happened to you last week (including your thoughts) by using no words—only pictures drawn on paper that look unambiguously like what they describe.

Of more recent vintage, language-based systems have been used by humans for the last several thousand years or so.[8] At some point in our history, people made the remarkable discovery that it was possible to use abstract markings to visually capture the sounds and words of speech. This discovery was epic, for it in essence led to the creation of a new and dynamic way to communicate, a way that could do much of what speech could, even doing some things far better. With the advent of language writing, speech at long last acquired a teammate of similar talent, marking a watershed achievement in human history.

Mathematicians from time to time have made similar conceptual realizations about their own discipline. One of the more recent ones occurred throughout the 1800s when mathematicians realized that mathematics was way, way broader than they had before thought. So much broader in fact that the medium-sized lake that they thought they were almost finished charting turned out surprisingly to be only a shallow bay in a very large, very deep, and very mysterious ocean—one literally teeming with life and waiting to be explored. More than a century and a half later, the mathematical community is still successfully investigating this massive ocean with no end in sight.

Now on to writing: What are some of the advantages that we gain by choosing to communicate this way? We give a few in the next section.

Some Advantages in Communicating by Visible Marks[9]

1. *Permanent communication*: Writing gives one the ability to communicate over great distances both in space and time. The mathematician Euclid has been dead for more than two thousand years, yet his landmark work *The Elements* remains and may still be used to instruct.
   A. *History*: The subject was immensely enriched with the invention of writing. A myriad of clay tablets with writing on them have been found in the Fertile Crescent giving us a glimpse into the world of peoples from long ago and also a better understanding of ourselves.[10]
   B. *Societal accounting*: Who owns what and how much of it do they have. A permanent record can be created and used in the case of disputes. Laws and contracts can be recorded as well as births, deaths, and other events. We regularly use receipts to demonstrate proof of purchase.

2. *Stationary, visible communication*: Things that are presented in writing are right there in front of us—allowing us to see, then pause, reread, study, and reflect upon them. These are evidently important features because now with recording devices (e.g., tape and digital recorders) we can make the actual sounds of speech permanent, but writing still is in no danger of extinction.[11] Writing gives us something that even the technological captures of speech don't possess; that something appears to include its static (think of still photographs versus video recordings) and visible features. Some linguists proclaim that these features have allowed writing to "entirely reshape our understanding of the world."[12] Ditto for the visible symbols of mathematics.

3. *Communication with oneself*: Writing offers great opportunities for self-communication.[13]

   A. *Personal convenience*: If you need to buy fifty items, rather than committing them to memory and taking the chance that the random sneeze of a stranger could cause the mental list to vanish, a printed list can be created on paper or some other material. You can go days without thinking about the items and once at the store you may reference the list with perfect recall. We regularly use writing as a memory aid when we jot down phone numbers, appointments, and e-mail addresses.

   B. *Recording one's thoughts, ideas, and data*: With writing one can keep journals or diaries that capture one's state of mind at a particular moment in time. One can also capture ideas and data for later use. This turns out to be extremely useful in the sciences. In fact, prominent figures of the scientific revolution, such as Francis Bacon and Galileo, recognized the critical importance to the entire enterprise of writing down information obtained from observations and then translating them into written mathematical form.[14]

4. *A tangible model for speech*: A printed model of speech allows for an analysis of the subject in ways that would not be possible with speech alone. For example, we can take statements in writing and experiment with them by moving parts of the sentence around (e.g., trying out different effects on meaning and impact by moving around or interchanging different adverbs or adverbial phrases). While it is certainly possible to do some of this in speech alone, it is worthwhile noting that often public speakers will first pen their speech in writing, evidently to gain a clarity that speech alone fails to give them. In a sense, writing gives to the analysis of speech what a map gives to the analysis of a geographical region.[15] Using written numerals to solve problems serves a similar function for situations involving quantity.

Some Disadvantages of Communicating by Visible Marks

Everything has side effects; some of these for writing include:

1. *Lack of context and immediate feedback*: When two people are talking to each other they have the luxury of knowing the context and are able to obtain immediate feedback from each other. Contextual awareness and instantaneous feedback often wind up being crucial components in the exchange.[16] Communication in writing is more impersonal and doesn't naturally possess either of these components. These issues become extremely severe in written mathematical communication.
2. *Literacy issues*: We acquire speech naturally; nearly everyone who is capable of speaking does so. On the other hand, acquisition of an artificial written symbolic code is not something we do naturally—even when it is a model for the spoken word (and much more so when it doesn't even have this advantage, as is the case with the symbolic expression of mathematics). Writing requires much more formal instruction. As such, a "literacy gap" can arise between those who learn to use the written symbolic system and those who do not. The measure of a person's literacy depends to some degree on the extent to which he understands and can use the written code.[17] A society, of course, can do a lot to eliminate those gaps through education.

The Alphabetic Principle

A particular aspect of writing that will prove useful later is vividly on display in alphabetic writing systems (such as written English).[18] All English words in writing are primarily constructed from a twenty-six letter alphabet (making allowances for punctuation marks, capital letters, etc.). This set of letters is closed in the sense that new symbols aren't added to it (over short periods, at least). When we encounter a new concept we want to represent as a word, rather than creating new symbols to describe it, we form instead a new arrangement or spelling from the same twenty-six letters.[19] Think about it, we can describe in writing much of the complexity in our world by using only twenty-six letters.

A similar principle is used in chemistry, where it is believed that the complexity of all naturally occurring substances is reducible to a basic set of fundamental naturally occurring chemicals called elements in varying combinations. These naturally occurring elements can be viewed as a closed alphabet of sorts, with newly discovered chemical compounds being written not as new elements but as new arrangements (or "spellings") of the closed set of basic elements. We will see this idea, of constructing more complex situations from a simpler basic set of building blocks, come up several more times in a mathematical context.

Naming Numbers

We have already encountered a few of the names for numbers that are in common use (i.e., ten, twenty-three, six hundred seventy-two, eighty-three, eighty-five, thousand). Many of the names used, however, were not necessary for the subsequent construction of the coin system that was introduced. Differences in size can be distinguished in writing without giving proper language names to each and every configuration of the visible symbols. The groupings for the various coin denominations hold whether we give names to them or not. For instance, without providing names for the resulting coin arrangements, we can rewrite the following:

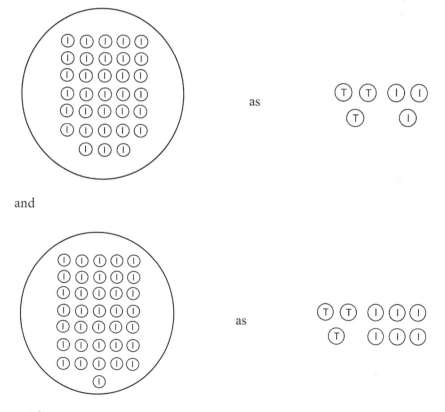

and

After grouping, we can clearly tell which of the two sizes represented symbolically is larger without ever attaching proper names to either coin arrangement. That is, even though we created and gave names to the each new coin denomination (ten, hundred, thousand, etc.), it is not necessary to give a name to every single combination involving them.

However, since it is people who, in the first place, introduced systems such as our coin numerals to help them better handle quantity and who primarily communicate through speech and language writing, it is also people who desire to give actual language names to each of these coin configurations.

To systematically give names to each arrangement, we must start by supplying names to the primary ungrouped configurations (remember any collection of tally marks that is less than ten strong is left as is):

$$①, ①①, ①①①, ①①①①, \ldots \ldots, \begin{matrix} ①①① \\ ①①① \end{matrix}, \begin{matrix} ①①① \\ ①①① \\ ① \end{matrix}$$

The common English names for those numbers are, respectively:

one, two, three, four, five, six, seven, eight, nine

These names can also be used to describe sequences involving the larger coin denominations. Just as we would describe AAA as "three-As," we can describe ⓉⓉⓉ as three-tens. This allows us to name the sequence of tens:

$$Ⓣ, ⓉⓉ, ⓉⓉⓉ, ⓉⓉⓉⓉ, \ldots \ldots, \begin{matrix} ⓉⓉⓉ \\ ⓉⓉⓉ \end{matrix}, \begin{matrix} ⓉⓉⓉ \\ ⓉⓉⓉ \\ Ⓣ \end{matrix}$$

by the compound words, respectively:

one-ten, two-tens, three-tens, four-tens, five-tens,
six-tens, seven-tens, eight-tens, nine-tens

We can do the same for sequences of higher denomination coins as well. Using these rules we can give the coin numerals the following names:

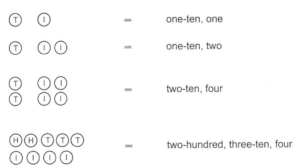

In actual practice, the following abbreviations are employed in English:

one-ten, one	≡	eleven	two-ten	≡	twenty
one-ten, two	≡	twelve	three-ten	≡	thirty
one-ten, three	≡	thirteen	four-ten	≡	forty
one-ten, four	≡	fourteen	five-ten	≡	fifty
one-ten, five	≡	fifteen	six-ten	≡	sixty
one-ten, six	≡	sixteen	seven-ten	≡	seventy
one-ten, seven	≡	seventeen	eight-ten	≡	eighty
one-ten, eight	≡	eighteen	nine-ten	≡	ninety
one-ten, nine	≡	nineteen			

Using these abbreviations allows us to describe these coin arrangements by their more common names, respectively: eleven, twelve, twenty-four, and two-hundred thirty-four.

Employing the names given earlier combined with the grouping names of ten, hundred, thousand, or a million means that every coin numeral designating size below ten million can now be communicated through speech or language writing as a word or compound word. We can easily extend this to represent larger whole numbers up to a billion and trillion and so on by introducing new coin denominations. Moreover, this situation is not unique to English; many other languages make use of similar conventions in developing words for numbers.

Order Counting

We now have three ways to communicate information about quantity—one way in speech (sounding out the language words described earlier) and two ways in writing (using either the coin numerals themselves or the written language words). The convergence of these three places within our reach a beautiful conceptual and computational jewel. Open wide to us now is a powerful newfangled way of counting.

Before discussing the details of this new manner of counting, let's first revisit our tally mark methods by using them to determine the size of our alphabet:

We first perform a tally of the letters in the English alphabet by tagging each letter with a tally mark:

A B C D E F G H I J K L M N O P Q R S T U V W X Y Z

I I I I I I I I I I I I I I I I I I I I I I I I I I

Grouping these in tens gives:

Changing denominations gives the total as:

 or twenty-six

In this example, we have determined the size of the alphabet through a method of tallying and grouping. Since we used it to determine the size of a collection, it qualifies as a type of counting. We will call it *tally counting.*

Next we will blend the properties of human language together with our coin numeral methods to fashion out an entirely new way of gauging the size of a collection.

Consider the named number sequence given below:

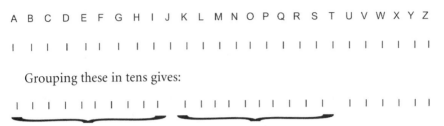

one   two   three   ••••   nine   ten   eleven   ••••

We can think of this sequence as a set of ordered cards and write:

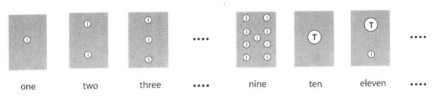

one   two   three   ••••   nine   ten   eleven   ••••

We will now demonstrate, with the English alphabet again, how using these cards gives us a way to count that is different in the details from the tally method of counting. In preparation of dealing the cards, we first reorder them as follows:

eleven   ten   nine   three   two   one

Lining up the letters and cards gives:

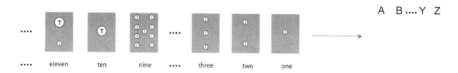

eleven   ten   nine   three   two   one   A B ....Y Z

If we deal the cards by passing out the "one" card to the letter A and the "two" card to the letter B and continuing on to the end of the alphabet, we get the following pairings:

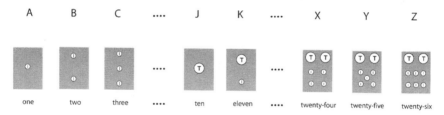

one   two   three   ....   ten   eleven   ....   twenty-four   twenty-five   twenty-six

The last value card dealt is the card with the value twenty-six. This value matches the size of the collection exactly, meaning that there are twenty-six letters in the alphabet.

The two methods of counting the alphabet yield the same result, but they are different in the details. In the first example, we perform the tally first and then apply grouping to obtain the value of twenty-six. That is, we still have work to do in obtaining an easy to understand answer after the initial tally mark assignments have been made.

In the second example, we tag the letters with coin configurations on cards that have already been ordered and grouped. This means that once the assignments have been made, no additional work is required—the last card dealt contains complete information as to the size of the collection (providing in one fell swoop, if you will, a complete memory of the tally). Moreover, now that each of the coin arrangements are named, the card assignments could just as easily have been the spoken words for:

one, two, three, . . . , ten, eleven, . . . , twenty-four, twenty-five, twenty-six

We can verbally sound out these words as we go through the alphabet—essentially giving us a way to determine the size of a collection by simply talking out loud (or silently if we prefer). For this method to work, however, the ordering of the cards or names must be perfect; as such, we will call this type of counting *order counting*. This is the method of counting that most of us first learned as children.

We now take one more look at how these two methods of counting work by giving a side-by-side example. We determine the number of names in the following collection using: the tally method and the order method.

Names = { Caspian, Superior, Victoria, Huron, Michigan,
Tanganyika, Baikal, Great Bear, Malawi, Great Slave, Erie,
Winnipeg, Ontario, Balkhash, Ladoga, Vostok}

The Tally Method

Tagging each of the names with a tally mark yields:

Caspian	Superior	Victoria	Huron	Michigan	Tanganyika	Baikal	Great Bear	Malawi	Great Slave	Erie	Winnipeg	Ontario	Balkhash	Ladoga	Vostok
I	I	I	I	I	I	I	I	I	I	I	I	I	I	I	I

Collecting the tally marks into groups of ten gives:

Changing denominations gives the total as:

$$\widehat{T} \quad \widehat{I} \quad \widehat{I} \quad \widehat{I} \quad \widehat{I} \quad \widehat{I} \quad \widehat{I}$$

There are sixteen names.

The Order Method

Tagging each of the names by dealing the ordered cards (including the language names for the numbers) yields:

Caspian	Superior	Victoria	Huron	Michigan	Tanganyika	Baikal	Great Bear
one	two	three	four	five	six	seven	eight

Malawi	Great Slave	Erie	Winnipeg	Ontario	Balkhash	Ladoga	Vostok
nine	ten	eleven	twelve	thirteen	fourteen	fifteen	sixteen

The last card dealt represents sixteen meaning there are sixteen names.

Notice the difference between the two methods: Nowhere in the tally method do we need the words or symbols for eleven, twelve, thirteen, fourteen, or fifteen, but these words are essential to determining the count for the order method.

Both methods have their advantages and disadvantages. The order method can be easily sounded out in speech, allowing us to count in a manner which, on the one hand, is quite sophisticated and powerful (containing within itself a complete memory of the tally), yet, on the other, is still simple enough to teach to preschool aged children; and if people are being counted, they can even aid in remembering the value and order of the count by calling out or memorizing their assigned number. The tally method is often preferred in the case of recording a running total. In practice, it is not uncommon to use the advantages of both methods in combination to measure size.

Number Names as Proper Names

The previous examples show that we can attach number names to the proper names for a group. There is nothing to prevent us from then actually using those number names to identify the objects of the group in the stead of their proper names. This is huge in circumstances where we care about order. Order occurs all around us from the way people arrive at an event, to the way pages are arranged in a book, to the order of countries according to population or test scores, to the way stations are configured on a television set, and so on. If we want to capture that order, number names are the best thing going.

Number names also can be used for identification. If there are millions of individuals that we want to uniquely identify, number names work much better than proper names since with the latter there is no predictable method to know how they are chosen, and there may be many people who have the same first, middle, and last name. Number names, on the other hand, are systematic, numerous, and far outnumber any collection that we might care to classify. Examples of number names as unique identifiers include social security numbers, driver's license numbers, student identification numbers, and so on.

Using numbers as opposed to names alone for street addresses illustrates how they can give us a navigational advantage. Consider the following two scenarios for a group of houses evenly spread out over a half mile in a neighborhood:

Using names as addresses:

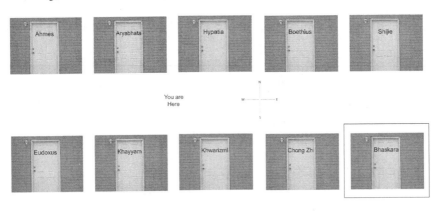

Using number names as addresses:

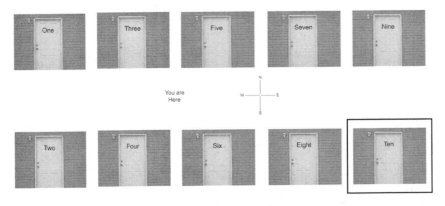

If a person, at the indicated location (in between the two rows of each addressing scheme), is looking for the house inside of the black square, then the system of addresses using numbers for names gives them more information. Using this system, after seeing the four adjacent addresses and knowing the predictable order of number names, she will know that the house she is looking for is to the east on the south side of the street. The proper name addressing scheme offers no such advantage since there is no predictable order that allows someone to find adjacent houses.

Using number names to order information in useful ways that provide for more convenient access is a practice that is entrenched deeply in the framework of our society.

Conclusion

Writing and speech offer two different yet complimentary ways to represent the world. We will find, when it comes to handling and communicating quantitative features of the world however, that doing mathematics with the aid of visible marks far outpaces doing mathematics verbally. The sizes of collections, in general, are so large and so varied that they are generally most effectively handled through using these more permanent and observable tools.

This means that many of the advantages and disadvantages of writing will also be inherited by the visible symbols we use to express mathematical ideas. We will also discover that the type of writing we will ultimately develop with numerals, while sharing much in common with language writing, will also have its own characteristics which make it a distinctly different form of communication as well.

Two critically relevant features inherited by mathematical writing are that it can make ideas visible and stationary. These properties often permit powerful and comprehensive new views of phenomena both inside and outside of mathematics. Such views allow for the easier recognition of patterns in the symbols or diagrams from which it becomes possible to build entirely new mathematical disciplines or to rephrase them in novel ways to make them available to a lot more people. These aspects will prove to be true game changers in helping to build an effective system of numeration for use by the average citizen.

The coin numerals that we have developed so far represent an important class of quantitative symbols—the class of additive numeral systems. These are systems in which you add up all of the individual numerals in a configuration to obtain the value represented. This is a very intuitive and natural thing to do, and while exceedingly useful (let there be no mistaking of this), these systems were eventually supplanted by far more potent numeral systems of the type that we use today. The emergence of this new way of representing numbers is nothing short of a heroic tale in scientific history. This book is, in part, a celebration of that marvelous episode in science, and of the known and unknown mathematicians who participated in it.

In the next chapter, we discuss how to transition from the coin system developed here to this more powerful and compact mode of representation. The gains obtained in this development will be a technological tour de force. In terms of importance, its discovery may very well be as the great French mathematician Henri Lebesgue says "perhaps the most important event in the history of science."[20]

3

An Ancient Tool Gives
Rise to Modern Mathematics

One is hard pressed to think of universal customs that man has successfully es-
tablished on earth. There is one, however, of which he can boast: the universal
adoption of the Hindu-Arabic numerals to record numbers. In this we perhaps
have man's unique worldwide victory of an idea.

—Howard Eves, American mathematician
and historian of mathematics[1]

EVERYONE WHO KNOWS ARITHMETIC, regardless of tongue, understands the
meaning of the mathematical statement: 2 + 3 = 5. We all write it the
same way using the same symbols. We don't, however, all say it the same way;
with speakers of one language in many cases saying this statement radically
differently from speakers of another language. We don't see this sameness in
writing and difference in speech in regards to English versus French or English versus Korean.

For instance, in English, the way we say and write the statement, "That sure
is a tall building," is very different from the manner in which someone who
speaks French or Korean would say or write it. Both the speaking and writing
of each of these two languages is drastically different from the speaking and
writing of English. Why is this not true of the statement 2 + 3 = 5? Why do we
pronounce it differently from the French or Koreans and yet write it the same
as they do?[2] To which language does this statement belong?

The circumstance of people the world over now writing their numerical
statements in practically the same way is a relatively recent phenomenon.
Throughout most of human history this was not the case at all; with the

number of known distinct numeral systems numbering more than one hundred.[3] Most people separated by great distances used, if they had them at all, different numerals as well. What happened to change this? There are still a wide variety of languages with an accompanying variety of different ways to write and pronounce numbers (one, un, uno, ein, один, moja, wahid, etc.), but no longer are there still a wide variety of numeral systems in everyday use. What is it that makes our present day numerals different?

Our aim in this chapter is to shed light on this interesting situation. At its heart is the issue of efficient calculation in writing. We want to do more with our numerals than just describe the sizes of collections; we also want to calculate with them. Are the twin requirements of communicating size in writing and effective calculation in writing compatible? We begin the investigation of this question by first considering addition—the simplest of the elementary operations of arithmetic.

Addition

Two very natural things we can do to a collection of objects is to add more objects or take some objects away. Let's start with adding objects to a collection, and to simplify matters for now, let's assume that the objects are cars. Just imagine for a minute that you collect cars and want to add more vehicles to your collection. To fully understand just how big your car collection will be, you have to take into account three numbers:

N1: The number of cars in the collection.
N2: The number of cars added to the collection.
N3: The total number of cars in the collection after the addition.

You can, of course, find the total number by doing a direct count after you've bought all of the cars but your goal is to avoid this. The plan is to calculate N3 directly from the symbols involved in representing N1 and N2. If you can't do this, then our ability to do arithmetic in symbolic form is going to be severely limited. We will be able to use the symbols for representation of numbers but not for calculation. We presently have two ways to represent quantity in writing—using coin numerals or using language number words. Let's look at how addition plays out with both.

Add fifteen cars to twelve cars using coin configurations. Symbolically, we have:

Un-Grouped Representation

Grouped Representation

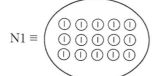

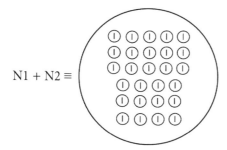

We first add using the ungrouped version and then group:

Grouping these in tens yields:

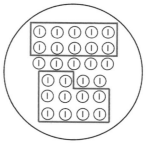

which is equivalent to:

(shading indicates changes in denomination)
Now we add using the grouped representation:

$$N1 + N2 \equiv \text{Ⓣ} \quad \text{Ⓘ Ⓘ Ⓘ Ⓘ Ⓘ} \quad + \quad \text{Ⓣ} \quad \text{Ⓘ Ⓘ}$$

$$\equiv \text{Ⓣ Ⓣ} \quad \text{Ⓘ Ⓘ Ⓘ Ⓘ Ⓘ Ⓘ Ⓘ}$$

You have twenty-seven cars in both cases.

We can add in grouped form and obtain the same answer as adding in ungrouped form. This turns out to be true in general, meaning that once we have written N1 and N2 in grouped form, we can find N3 by staying completely within this grouped form. We don't ever have to return to the ungrouped form and thus are able to use the full force of grouping to perform the operation of addition.

In this first example, we successfully performed the addition using coin numerals and nothing else. Can we do the same with the language words for numbers? That is, can we successfully add fifteen and twelve using only the letters in each word with no assistance from numerals?

Let's try it:

Symbolically we have N1 ≡ *fifteen* and N2 ≡ *twelve.*

Thus: N1 + N2 = *fifteen* + *twelve* = ??

Unlike with coin numerals, there is no obvious way to construct the answer to this sum using only rearrangements from the thirteen letters in these two words. The following are three possible rearrangements from the thirteen letters:

Arrangement 1: *fifteentwelve*
Arrangement 2: *fiftweteenlve*
Arrangement 3: *ffiwfetleveen*

None of these, or the millions of other rearrangements possible, works. If, however, we allow in our findings with coin numerals, we know that the answer is twenty-seven. In language words, we then have:

fifteen + *twelve* = *twenty-seven*

It is not possible, without such assistance, to assemble the answer to this sum from the initial components (letters) of *fifteen* and *twelve*. Simply, combining the letters in these words will not construct the "*y*" and "*s*" in the word *twenty-seven* no matter what we do (nor make the *l*, *i*, and two *f*s disappear either). In fact, rearranging the letters in these two language words not only leads to nothing new in mathematical content it also leads to millions of rearrangements that are devoid of language content as well.

Language words are composed out of letters which serve the purpose of conveying in part the sound content of speech, not the mathematical content of quantity. This means that these words and letters by themselves do not contain within themselves the seeds of their own solution—they don't have a natural syntax for combining quantity, if you will.

Coin numerals, on the other hand, do contain within themselves the seeds of their own solution—that is, they possess a natural syntax for dealing with quantity. This is precisely why they were built. This yields a straightforward way to build the answer Ⓣ Ⓣ   ① ① ① ① from the constituent parts: ① ① ①

Ⓣ ① ① ① ① ① and Ⓣ ① ① (you just combine them together while still retaining meaning).

It appears then, that with respect to adding values in writing, coin numerals have a decisive advantage over language words. This is true in general and we will act accordingly; reserving the right, however, to use the language words to represent numbers in the situations for which they were designed, such as basic communication in English. The next example shows how to handle additions involving multiple changes in coin denominations.

Let's add seventy-seven and sixty-five using coin numerals:

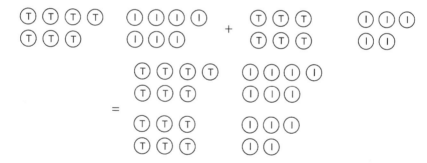

Organizing the coins in groups of tens yields:

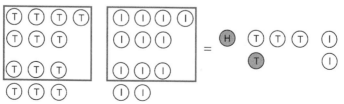

or one hundred forty-two.

The practice of performing additional grouping after the initial addition is equivalent to the process known as carrying.

It is important to note that even if writing media and instruments are not readily available, we could still create a workable grouping scheme by fashioning out dozens of material tokens representing each type of coin denomination and physically carrying out the arithmetic. Or we can form a system out of some other sorts of physical objects like the one given here:

Denomination	Object[a]
One	Pebble
Ten	Blade of Grass
Hundred	Kernel of Corn
Thousand	Oak Leaf
Ten Thousand	Stick
Hundred Thousand	Acorn
Million	Big Rock

[a]One blade of grass ≡ ten pebbles
One kernel of corn ≡ ten blades of grass, etc.

The representation and addition of numbers would work here just as they do in coins. After all, the system using pebbles, blades of grass, or other objects is simply a different version of the same fundamental idea of representing objects by symbols and then grouping them in denominations to make them more comprehensible, much like spoken French and spoken English are different versions of the same fundamental idea of communicating by sound.

So if we had enough of each item, it is possible to perform addition with sticks and stones (mimicking the additions we can do in writing with coin numerals). You can choose your own physical grouping system simply by replacing the pebbles, blades of grass, sticks, and so on, by another set of seven objects (e.g., peanuts, pieces of paper, spoons, glasses).

The Abacus

Basic additions, such as those performed earlier, are certainly workable using written coin numerals. It is a very straightforward procedure. *But what if we have to do a lot of additions?* Imagine how much work would be involved if we had to perform eight-hundred additions like the one we just did in finding the answer to "seventy-seven plus sixty-five." This is a very serious concern, for as we will see in our later discussions, there are many situations in life that require lots of additions for their solution. What are we to do about this? What did the ancients do?

Throughout history, societies around the world were faced with the problem of how to effectively handle a large number of additions. What is interesting is that many of them, often independently of one another, chose to deal with it in a similar way. What they did was develop an entirely different way to represent numbers, not in writing but by way of a mechanical device that sometimes involved the use of a frame and rods. The heart and soul of the scheme was to use position or location to indicate the size of the number as opposed to using coin denominations. Beads or other objects were used to indicate how much of each coin was present. This device or counting frame was often called an *abacus*. It was a devastatingly powerful idea.

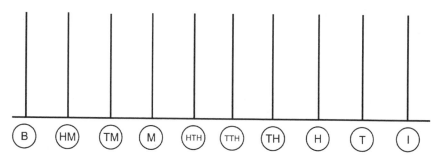

The coins in each column were not generally present in the physical devices and have been included here to facilitate understanding. Let's see how it works.

Here we see how four (H), seven (T), and six (I) coins (four-hundred, seventy-six) get represented in different columns on the abacus.

Coins   Abacus

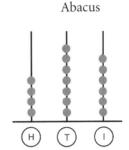

Our grouping rules are still in effect, so if we have ten beads in one column, it can always be replaced by a single bead in the adjacent column to the left. Thus,

is equivalent to

In coins, this reads as:

is equivalent to

We can also perform additions with this device by adding three hundred sixty-seven to four hundred fifty-six. The steps are broken down into individual components and the coins are included to aid understanding.

Adding with coins:

Adding on the abacus:

Combining coins and changing denominations:

On the abacus place the beads on each of the hundreds columns together (then doing the same to the tens and ones columns) and then carrying over to a higher denomination, where required yields:

The answer is eight hundred twenty-three.

The story then, for many societies, has been to develop a written symbolism (such as our coin numerals) for the representation of numbers. Unfortunately, this representation rarely allowed for quick, convenient, and easy-to-learn calculations (particularly in the case of multiplication and division). As compensation, a calculating device of some sort such as the abacus was employed.

The Abacus in History

The abacus dates back to antiquity. One of the oldest physical artifacts of the device is a Roman hand abacus from ancient Rome. It is also believed that abacus-like devices were used in Mesopotamia as early as the third millennium BCE.[4] Use of an abacus, at some point in their history has been documented in many other places, including China, Japan, Greece, Russia, and Central America. Some of these devices are still in use today. Below is a real world example of the fundamental idea of representing value by location.

Chinese Abacus or Suan Pan (ca. 1500–1600 CE)[5]

In the following picture, the beads in the top section (sometimes called heaven) have value five while the beads in the lower section (sometimes called earth) have value one. The beads are active when placed toward the center divider. A value of up to fifteen can be represented on a given column. None of the beads are active in this picture.

Photo taken by author.

Placing the coins below each column shows that we can represent numbers into the billions on this particular device.

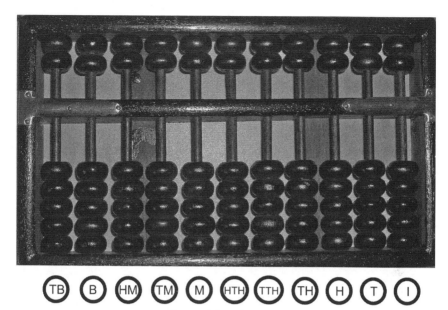

Photo and design by author.

The next diagram shows how to represent nine million, six hundred three thousand, one hundred fifty-four on the Suan Pan (note that the gray beads are active).

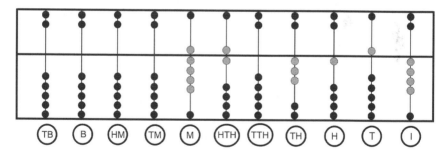

The theoretical device introduced at the beginning of our discussion on the abacus differs from the Suan Pan in a fundamental way—its beads can be either created or destroyed—with the rule being that visible beads are active and deactivated beads disappear. On a physical abacus the beads can't simply vanish, so there has to be some other means, such as using a center divider, of distinguishing when they are active or inactive.

Elementary Arithmetic Has Hidden Complications

The difficulties inherent in performing lots of additions or lots of subtractions with coins lays bare the fact that elementary arithmetic is harder to rein in than first appears on the surface. We have already seen that the English language words for numbers are not well suited for performing even the simplest of additions, and coin numerals, while much better at addition than words, also have their limitations.

Thus in arithmetic, you may easily devise a written scheme by which to represent or describe numbers, but when you start to perform addition, subtraction, multiplication, and division within this scheme, it can be a major pain in the neck obtaining answers. This is not a trivial problem, and it was the abacus or some other such device that often offered a way out of the difficulty.

Consider a large collection of files, say ten thousand, which are identified by name. The major issue of concern with such a collection is how best to store files so that information can be obtained efficiently. There are a wide variety of options, all of which accomplish the task of storing and keeping the records safe. However, not all of these options lead to quick retrieval. For example, a random nonalphabetical storage keeps the records secure but in general will not lead to the efficient retrieval of a given file. Files certainly can be accessed, but it may take quite a while depending on their location. However, if we store them alphabetically based on last names, retrieval can take place more rapidly. The point being, that if one wants to be able to retrieve files quickly, one must pay attention to how one stores those files.

In a similar fashion, if one wants to calculate efficiently within a written symbolic system, one must be much more careful in the construction of such a system than one might initially think. Most written systems for representing numbers, including our coin system, were handy for describing the varying sizes of collections and distinguishing among them, but were not so useful for quick, convenient calculation.

By using an abacus one could translate the numerals from any written system to the device and perform calculations far quicker than possible within the written system—the calculation reduces to a mechanical procedure involving the sliding of beads (or some other physical manipulation) as opposed to the generally slower procedure of writing down a large number of symbols. Once the answer is obtained it could then be translated back to the original numerals.

While using a device to help one do calculations is not in and of itself a bad thing (today we do so with electronic calculators), it is, like the spoken word, a transient process. Once the calculation is done, all of the intermediate steps are lost. In some cases, for example, in education or bookkeeping, the loss of these intermediate steps can cause unnecessary difficulties. Moreover, effective use of the abacus is a highly involved process requiring much memorization, a great deal of understanding, and good hand-eye coordination. These were the skills demanded, not of a layman, but of a craftsman, and as with most crafts it took time and practice to master the art.

So while using a two-tiered scheme, representation of numbers in writing and calculating with numbers on the abacus is a workable scheme, it is not necessarily the most desirable state of affairs.

This brings us to a high point in our journey: Sprawled out before us now, on the one hand, is the technique that allows for the permanent and visible representation of numbers as numerals while, on the other, sits the technique of efficient and quick calculation with the abacus. The advantages and disadvantages of each are clear. Systems like our coin numerals allow us to record, visualize, and study quantity at our leisure but don't provide for swift and efficient calculation, whereas devices like the abacus, while allowing for rapid calculation, are difficult to learn and fail miserably at providing permanence and static visibility. We want the best of both worlds. Can we get it?

The answer is a resounding yes, and how it was accomplished comprises one of the great, drawn out tales in mathematical history. The specific details of exactly who and exactly when are shrouded in mystery, but what is known is that the fulfillment of this dream came in the guise of a special system, a system with panache, complete with its own brand and whose arrival from the East was destined to change the world.

Script of the Abacus

We have already seen that powerful new ways to represent the world were developed by representing huge portions of speech with visible marks. Similarly, we will now see that some of the most powerful ways to represent quantity in writing come from an attempt to capture, in a system of numerals, the processes at play on the abacus. The stage has already been set with our graphical representations of the abacus. We now follow this path through its "natural" course.

The great strength of the abacus lies in the fact that it is symmetric in the rods or columns. That is, the following bead design

can represent three, thirty, three hundred, or three thousand, depending on its position of the device. In contrast, representing these values in coins would require that we use four different denominations.

three

thirty

three hundred

three thousand

The symmetry of the abacus means that what can happen on one of the rods is what can happen on any of them. Thus although the "values represented" change when we change positions, all of the content possible on any of the rods can be learned by simply studying just one of them. This means that if the number of distinct designs on one of the rods of the abacus is finite, then the number of distinct designs on any rod of the abacus is finite, and represents a self-contained set. This will afford us the magnificent opportunity to inject the alphabetic principle into our numeration scheme.

As an example, we can represent the design corresponding to three beads on a rod, |, by the symbol ⭐. Using this allows us to rewrite the entries as:

three

thirty

three hundred

three thousand

The representation for three thousand three hundred thirty-three on the abacus translates as follows:

becomes

The number of designs that can occur on any rod of the abacus is indeed finite due to our grouping rules. The following list gives the ten different designs that can occur on the ones rod of an abacus (we will refer to these as the "ten original designs").

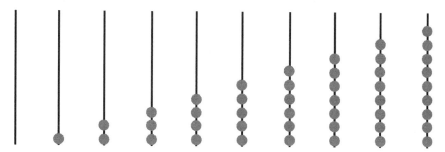

These are also the only designs that can occur on the tens rod as well (note arrows).

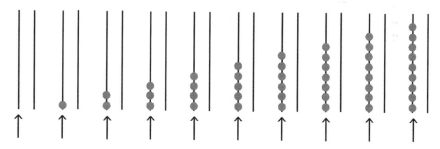

The designs with ten or more beads are not listed since they will correspond to a repositioning of the beads based on our grouping rules, and we will then have multiple rods, each respectively matching one of the "ten original designs" represented on the ones rod. For example, if we have twelve beads on the ones rod this reduces as follows:

Each one of the designs on the two rods on the right is listed in the "ten original designs" (think of the two rods as a two "letter" word from the "ten-letter alphabet" given by the "ten original designs" where we think of each design as a "letter"). If those twelve beads happen to be on the tens rods, instead, then they transform as follows:

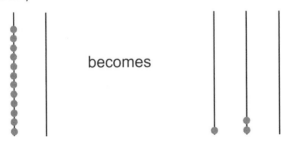

Once again each one of the designs on the three rods on the right is represented in the "ten original designs" (think of the three rods as a three "letter" word). We can always perform carries like this when we have more than nine beads—so the number of different scenarios that can occur on the ones or tens rods are exactly those given by the "ten original designs." These arguments translate directly to the hundreds rods, the thousands rods, and so on (with the only modification being that we get longer words from the "ten original designs" thought of as a "ten letter alphabet"); which means that the "ten original designs" are the only ones that can occur on any rod of the abacus.

If we consider the following four rod abacus, | | | | , any number between one and nine thousand nine hundred ninety-nine can be represented by some combination involving only the "ten original designs" (the words to describe these situations will range in length from one "design" long to four "designs" long). For example, all four of the rods in three thousand four hundred seventy-three come from the designs listed in "ten original designs" (duplications allowed of course):

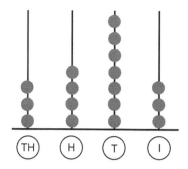

Given that any number in the stated range can be represented by some combination of the "ten original designs," we will gain great advantage if we replace each original design by a single symbol as we did before when we re-

placed the ⁞ design by the symbol ★ . This time, however, we will use our modern symbols to name the "ten original designs" and we have:

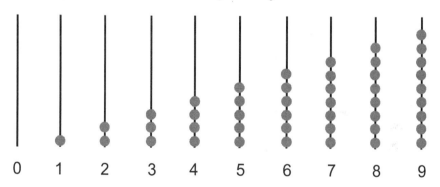

These symbols are called the Hindu-Arabic (HA) numerals. The Hindu part of the name comes from the prevailing evidence which points to this numeral system having originated in ancient India (most likely somewhere between the first and seventh centuries CE). The Arabic part comes from the fact that the numerals were disseminated to Europe and eventually to the rest of the world through the Arabs in the Middle Ages (ca. 800–1200 CE).

Notice that, unlike the beads, these symbols of and by themselves give no clue as to the number they stand for—they yield a coded representation.[6] As with words in a language, an investment in memorization must be made. In what follows, we will see that this investment is one of the most valuable a person can make. Using these symbols for their respective situations, the previous representation of three thousand four hundred seventy-three becomes:

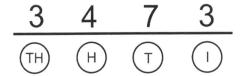

The HA numerals already hint at their potency by offering us a more compact way to represent numbers. Let's now see how they compare with coin numerals and the abacus in representing the number six million, two hundred sixty-two thousand, five hundred nine.

Coins:

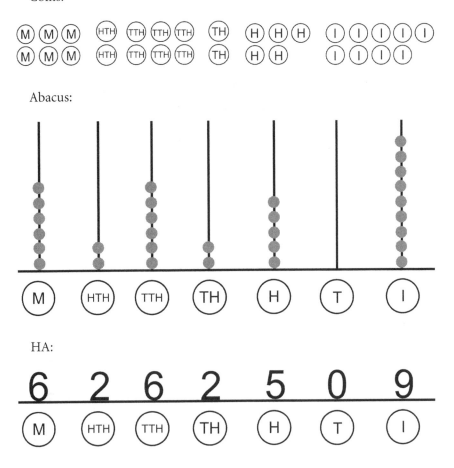

Abacus:

HA:

The coin representation contains thirty total symbols in six varieties. The abacus contains thirty total symbols all of the same variety but in different columns. The compact system using HA numerals only requires seven symbols.

The abacus in practice does not have the coin tags below the columns. A small investment in memory of the various place values allows us to work the device without them. Similarly, memorization of the place values allow us to dispense with the tags in the case of the HA symbols as well. So in the example above, we can more simply express the number as 6262509, with no loss of information.

Below is a comparison of the abacus rods and HA numerals without coin tags:

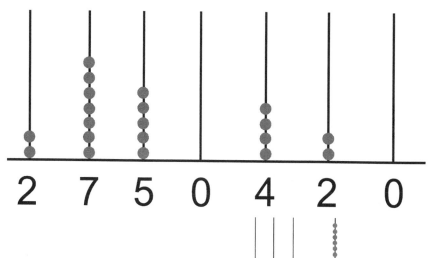

We see then that the ten distinct designs {|, |, |, ..., |} that can occur on any rod (ones column, tens column, hundreds column, etc.) of the abacus developed into the ten distinct symbols {0, 1, 2, . . . , 9} that can occur in any place value (ones, tens, hundreds, etc.) of a string of HA numerals—which puts us in the familiar position of once again being faced with the prospects of an alphabetic situation.

In this new script, any whole number no matter its size can now be expressed by simply using the ten symbols {0, 1, 2, 3, 4, 5, 6, 7, 8, 9} in some combination. This fundamental set of ten digits is self-contained just as an alphabet is in language. It does differ from a language alphabet in one crucial aspect however: Random combinations of the ten digits will still lead to valid numeral words whereas arbitrary combinations of letters will not generally lead to language words, unless it is by just plain luck.

In English, mistakenly writing "beelieve" for the word "believe," or "mitake" for "mistake" causes no problems in comprehension. Both spellings are close, and there is really no other word that the "mispellings" could be. This built-in redundancy in English is what makes spell-checkers possible in word processors.[7]

However, mistakenly misspelling "19900" for the intended "1900," by including an extra "9," is an extremely serious error in HA script, since unlike the situation in English, both strings are valid and distinct expressions. How do we know which is correct? We don't have pronunciation cues to guide us as in language and if there is no context to work with then we have a real problem. This was one of the major reasons for resistance to the adoption of the HA script—leading to their being banned in Florence, Italy, in 1299.[8] It remains a valid concern today which we still address when we write the dollar amount on a check both in digits as well as in language words. The words serve as a confirmation of the digit amount.

It is interesting to note that a misspelling of this type is less likely to occur in our coin system. Writing these numbers as coins yields:

19900 = (TTH) (TH)(TH)(TH)(TH)(TH) (TH)(TH)(TH)(TH) (H)(H)(H)(H)(H) (H)(H)(H)(H)   &   1900 = (TH) (H)(H)(H)(H)(H) (H)(H)(H)(H)

Note that "19900" in coins contains nine more symbols than "1900" making it highly unlikely that a misspelling off by nine symbols would occur in this script unless it was deliberate. This illustrates the very important phenomenon that issues that arise in one system may present little or no difficulties in another one. The mathematical and scientific landscapes are full of examples of this. The fact that computers are written using a system of numerals involving only 1s and 0s is a case in point. The circuitry involved with most computers deal with only two states—the presence of a voltage and the absence thereof. These situations are much better matched by a base-two (binary) system which uses only two symbols (0 and 1) than they are by one which uses the ten symbols (0, 1, 2, . . . , 9).

Is Zero a Number?

Our efforts to capture the abacus in written form have created an interesting and unique situation. Looking back to the original conditions involving the enumeration of cattle with tally marks, the smallest situation we had to consider, in order to measure the number of cows a rancher has, contains a single mark. Whenever we initiate a count we always have at least one element; otherwise, there is nothing to do.

However, our quest to make representation and calculation simultaneously convenient and also written led us to consider the "ten original designs" that can occur on the ones rod of an abacus. One of those designs is the rod with no beads. The other nine designs clearly represent numbers and have counterparts in the system of coin numerals:

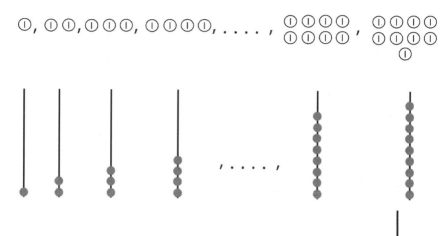

There is no room, however, in the coin system for the vacant rod: |. What should we consider as its counterpart?

One thing is certain, the empty rod and the HA symbol representing it are absolutely essential to the workings of the abacus and our written symbolic representation of it. We simply cannot do without them. For example, representing four thousand two on the abacus requires that the rods for the tens and hundreds be vacant. Using the HA script without zero yields:

$$\frac{4}{\text{(TH)}} \quad \frac{\quad}{\text{(H)}} \quad \frac{\quad}{\text{(T)}} \quad \frac{2}{\text{(O)}}.$$

However, if we remove the place value coins, we have 4   2 and it is hard to interpret exactly whether this is forty-two, four hundred two, or four thousand two. On the other hand, if we include the 0s, we have

$$\frac{4}{\text{(TH)}} \quad \frac{0}{\text{(H)}} \quad \frac{0}{\text{(T)}} \quad \frac{2}{\text{(O)}}$$

and we can safely remove the place value coins and obtain 4002 with no loss of information.

In systems such as the HA system, where we use the same symbols in different positions to denote value, we must keep track of what is absent as well as what is present—making the symbol 0 essential in these systems. In systems, such as our coin numerals, which do not incorporate position using instead different symbols to denote different magnitudes, a counterpart for the symbol 0 is not needed since it is clear what symbols are present and what symbols are not in the representation of a number. For example, in the coin representation of four thousand two we would simply write: (TH) (TH) (TH) (TH) (I) (I). No symbol is needed to denote the absence of the ten and hundred coins—their nonappearance is enough all by itself.

What then is the symbol 0 really? Some notational systems seem to require its use whereas others do not. Does it represent a true and bona fide number in its own right or is it simply a device to account for the absence of a place value in a positional system?

Historically, zero has been viewed both ways. It was first used as a convenient accounting device or placeholder in positional systems (much like how spaces separate words in a line of text). The Mesopotamians (ca. fourth century BCE), the Chinese (ca. 700 CE), and the Maya (a few centuries after the Chinese) all originally used zero in this manner.[9]

It was secondly viewed as a number in its own right. The first known recorded use of zero being considered as a stand-alone number was by the Indian mathematician Brahmagupta in his work *Brahmasphutasiddhanta* (ca. 628 CE).[10] This is how we view zero today. This viewpoint has led us to discover and utilize properties of zero that no other number possesses. It is one of the most important numbers in all of mathematics and plays a crucial role throughout the whole of mathematics (e.g., solving equations in algebra, simplifying expressions, as an identity element).

It is interesting to note that numerals employing negative signs were also employed as convenient accounting mechanisms in business (as debits) well before they gained widespread acceptance among mathematicians as being representatives of a new class of true numbers (negative numbers).

A Discovery within Mathematics Itself

There are two major points to discuss here. First, zero does not appear in our initial models for counting collections but rather appears as a necessary punctuation symbol to avoid ambiguity in a place-value system. As such, it is a number that was not discovered from the natural analysis of a physical situation but rather from formatting considerations within a given symbolic system. It simply pops up out of our reasoning.

All of this would seem to indicate that discoveries may happen not only from trying to symbolically represent what we observe in the physical world, but may also occur from trying to better represent and make more efficient what we observe in the symbolic world of mathematics itself. That is, the symbolic systems which we create and the rules governing them are to some extent worlds unto themselves and new discoveries are just as possible when trying to solve problems in these virtual worlds as they are in trying to solve problems in the physical world (and in many cases, given the convenience in working with symbols, the problems are easier to pose and treat in these virtual worlds). We observe similar scenarios happening time and again elsewhere in mathematics and in the physical sciences at all levels.

The conservation of energy law in physics is a case in point. One of the most powerful ideas in all of science, this law is based on the indirectly observed quantity of energy. In regards to the law, celebrated Nobel Laureate Richard Feynman says:

> It states that there is a certain quantity, which we call energy, that does not change in manifold changes which nature undergoes. That is a most abstract idea, because it is a mathematical principle; it says that there is a numerical quantity which does not change when something happens. It is not a description of a mechanism, or anything concrete; it is just a strange fact that we can calculate some number and when we finish watching nature go through her tricks and calculate the number again, it is the same.[11]

Second, although the two systems may appear to be equivalent versions, they can differ not only in convenience but also in range or extension. A symbolic scheme where value is based on position leads to an accounting symbol, 0, that ends up being the gateway to a new number called zero; whereas a simple grouping scheme such as our coin numeral system misses out on this number completely. Thus, although the two symbolic systems may have a common origin and common goals, hence sharing common features structurally, as we attempt to extend them or study them in greater detail, we find that they are never truly exactly the same.

Although they are identical in crucial respects they are also different in other respects and some of these differences can give one system a decided advantage in certain contexts. We saw this happen earlier, where the method of comparing collections involving lanes and the method involving small rocks had the common origin in comparing two herds of cows but with the small rocks we could also compare collections that the lanes method could not—things such as houses, trees, and other immovable objects.

Conclusion

We have now twice seen how powerful reorganization can be in arithmetic. In chapter 1, by reorganizing tally marks we were able to create our coin system of numeration (with denominations of ten, hundred, thousand, and so on) which greatly enhanced our way of expressing and comprehending larger numbers.

In this chapter, we have reorganized the coins into a positional system yielding even greater treasure. In the process, we have discovered a new number (zero) and also have, as we shall soon see, sown the seeds of a system that will allow the average person to perform with ease calculations that would have at one time taxed the skills of many an excellent mathematician. The simple concept of reorganization plays a major role in mathematics at all levels and in life in general.

Now that we have the compact and efficient HA numerals, the stage is set to apply them to adding, subtracting, multiplying, and dividing. Before doing this, however, we will take a detour to see how tinkering with symbols and diagrams empowered an ancient scientist, through reasoning and calculation, to determine the distance around what at the time was a very big and very mysterious earth. A value that was impossible for him to obtain directly by taking a measuring tape around the earth. What follows is a testament to the breathtaking power human beings are afforded when they employ reasoning and symbolic manipulation in clever conjunction.

4

The Ancients Perform
Miracles with Numbers

To see what is general in what is particular and what is permanent in what is transitory is the aim of scientific thought.

—Alfred North Whitehead, British mathematician,
logician, philosopher, and educator[1]

WE NOW ATTEMPT TO breathe life into an important chain of events from the distant past whose conceptual importance still resonates today. In our discussion, we will tell the tale of a fictional character, Amar, from ancient Sumer (the southern part of modern-day Iraq between the Tigris and Euphrates Rivers). And while his personage is imagined, Amar's story will not altogether be fantasy; for some person or persons had to actually discover the same things that he did. However, since we don't know the specifics about them, even their names, we have created Amar to represent them.

We will also share details from the life of the very real and very insightful third century BCE "Renaissance man" Eratosthenes who, through studied application of the work attributed here to Amar and others, was able to gauge the extent of the earth. This tale will provide dramatic demonstration to why human beings have cared about mathematics throughout the past and why they still care up to this very day.

Consider the game of chess: It is played on hundreds of thousands of individual boards with millions of individual pieces all around the world. If we want to learn how to successfully play on all chessboards, all that is required, remarkably, is that we learn how to successfully play on one of them. A person who becomes a true *grandmaster* on their chess set at home is a *grandmaster* on any chess set. All of the many game sets represent different forms upon which to play the same game, meaning that any particular chessboard can serve as a representative for all others. Thus in learning how to play on a particular representative, one is actually learning how to play on all representatives. Exploitation of this fascinating property of nature, in its more subtle forms, allows us to literally investigate the world from home.

Amar's Tale

More than 4,000 years ago in ancient Mesopotamia, a mathematician named Amar has decided to study circles. He sees circles all around him, both in the sky as well as on earth and is curious to learn more. Despite the fact that the circles differ in size, composition (string, rope, wood, dirt, imaginary paths, etc.), and location, Amar knows that they all have something structurally in common—that feature that allows him to recognize them all as circles.

To gain more knowledge about circles he realizes that he will need to make his own representative models and study these in great detail. He knows that he can easily construct a wide variety of circles by using nothing more than a pin, something to write with, something to write on (dirt, papyrus, soft clay, paper, etc.) and some string. For our purposes, let's assume he uses a stick for writing and dirt as his canvas. He constructs his circles by pinning down one end of a string and attaching the stick to the other end and tracing the figure in the dirt. Given that circles are so widespread, he realizes that there is a good chance that his pedestrian study of ones in the dirt will be representative forms of all circles—causing his results to have far-reaching implications.

It is evident from the very construction itself that, for each of his models, there is a point in the interior of each circle that is the same distance from every point on edge. He gives a name to this special point, calling it the "center," and also gives the name "radius" to the common distance. He also notices that if he goes from one side of each circle to the other side in a straight line through the center, the distance is always the same (no matter where he starts); moreover, its value is always equal to twice the radius. He calls this number the diameter for the given circle.

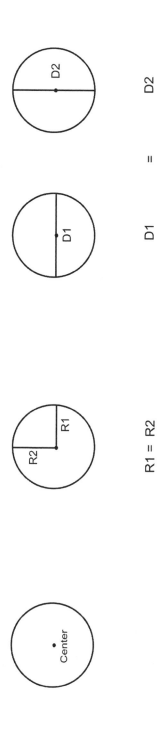

Every circle has a center

R1 = R2

All radii on a given circle are equal

D1 = D2

All diameters on a given circle are equal

At this point, he decides to see if other circles not of his making (e.g., wheels) also have a center, radius, and diameter. His analysis quickly reveals that regardless of size or material, all of the circles investigated do indeed have these three characteristics. In the midst of his investigations, he happens upon an accident involving an overturned cart and observes a piece of caked mud on one of the spinning upturned wheels. He imagines the mud rotating by itself in the air on the same path without the wheel and realizes that it would still trace out a circle.

In a flash of insight, he realizes that it is possible to link circles to turnings, where the whole circle corresponds to one complete turn, half a circle corresponds to half of a complete turn, and so on. He decides to create a unit of measure, which he calls the degree, to more finely gauge the amount of turning. In his system, 360 of these degrees correspond to a complete turn, 180 degrees to a half turn, 90 degrees to a quarter turn, and so on.

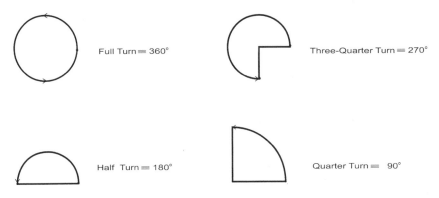

Full Turn ≡ 360°   Three-Quarter Turn ≡ 270°

Half Turn ≡ 180°   Quarter Turn ≡ 90°

He records this information and then sets about tackling the job of devising methods to calculate the distances around his circles (i.e., the perimeter or circumference of the circle). He can readily do this for his small circles by simply taking a long enough piece of string and wrapping it around the loop—then straightening it out and measuring the length. This method is general—meaning that it will work for any circle given enough string. It is not convenient, however, to use on very large circles. He decides to search for other methods.

Noticing that the larger the diameter of the circle, the greater the distance around it, he muses that there must be some relationship between the two. His tests reveal that the distance around his circles can closely be approximated by simply multiplying the diameter by 3.13 (this value was derived from the fraction $\frac{25}{8}$ which is the value that the early Mesopotamians reportedly used as an estimate for Pi; the better three digit approximation today is 3.14).[2] This number comes up every time, regardless of the size or type of circle. Why this strange number (and not something nice like 2 or 5) occurs and whether it comes up for all circles are mysteries to him. Nevertheless, he conjectures that it is true for all circles and catalogs this information in the formula:

Perimeter or Distance around the circle $\approx 3.13 \times (\text{diameter})^3$

If he uses this formula he can more than triple the size of the circles he can measure using a given amount of string. For example, if he only has 200 feet of string, then the wrap-around method will allow him to measure circles whose perimeters are 200 feet or less. With the formula he can use the same string to measure diameters up to 200 feet which correspond to circles having perimeters (or circumferences) of up to 200×3.13 or 626 feet, a significant improvement.

Giddy with success, he continues on and finds out that he can incorporate his new degree measure into the discussion. What makes the diameter method work is the multiplier of 3.13. Once the diameter is found, all one needs to do is multiply it by 3.13. His degree measure offers a much greater multiplier. If he can find the distance on a circle that corresponds to 1 degree of turning, he can then multiply this value by 360 to obtain the distance that corresponds to 360 degrees of turning (i.e., the entire distance around the circle). Thus to measure the distance around the 626-foot circle would in theory require only 626 divided by 360, or about only 1.7 feet of string.

We illustrate how the degree method works by finding the distance around a circle in which 1 degree corresponds to 2 feet. (Note circle is not drawn to scale.)

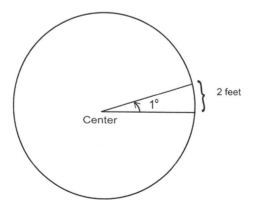

There are 360 sections (each 2 feet in length) just like the one here. We can find the distance around the circle by computing 360×2 feet, which equals 720 feet.

Amar's results compile as follows:

- Every circle has a center, radius, and diameter.
- Every circle can be partitioned into 360 equal units called degrees.
- The distance around a circle can be obtained or approximated in the following three ways:
  1. Encompass or trace out the perimeter of the circle with a long enough string, then straighten out the string and measure its length.
  2. Find the diameter of the circle and multiply this value by 3.13.
  3. Find the distance on the circle which corresponds to 1 degree, then multiply this value by 360.

He is fairly sure about all of his conclusions except for number 2. Although he has analyzed dozens of circles, he can't logically reason why the number 3.13 always gives a good approximation. It is an important enough aid in calculation that he must include it but he knows it will be up to others to give a definitive reason why it holds.

Amar has played his game well. The results discovered on the comparatively few circles investigated are indeed representative and apply to any circle (just like the rules and strategies of chess learned on one board being true on any chessboard). His efforts correspond to him first learning how to "play with circles" on his home game set and in the process learning how to "play with circles" on any set. His recording of these results means that humankind as a whole also learns how to "play with circles" on any set. The circle can be small enough to be written on a piece of paper or it can be one as big as the very earth itself. It simply does not matter.

Time passes and other fundamental discoveries are made, one being that the earth, though it appears flat, is in all likelihood probably round. One of the observations that lead to this conclusion comes from sailors noticing that when they approach a mountainous coastline, the first landmark they see is the top of a peak and not the entire mountain as you might expect if the earth were flat. As the ship gets closer and closer to land, more and more of the mass becomes visible. This same phenomenon is observed when a person walks up a hill with a mountain in the distance: He sees the summit first and then more and more of the prominence as he approaches the top of the hill. Since there are no hills in the ocean, a logical explanation for this occurring at sea is that the surface of the earth is curved.

Eratosthenes of Alexandria

The time is now the mid to late third century BCE and the great library in Alexandria, Egypt, is in its heyday. It contains much of what ancient man knows about the world. Among its holdings are papyrus scrolls which include the results of Amar's work on circles from so long ago. Its librarian is the great scientist and Renaissance man Eratosthenes. History will be much kinder to him than were many of his contemporaries, to whom he was known as jack of all trades but master of none.[4]

This exceptional man has tasked himself with the problem of determining the distance around the earth. Many scholars have come to believe that because the earth is round it must indeed be spherical in shape. Eratosthenes knows that spheres have circles in abundance (e.g., the equator, lines of longitude). He feels certain he can bring to bear humankind's knowledge about circles to solve this problem.

He knows about Amar's three ways to find the distance around a circle and concludes that number 1 and 2 are well nigh impossible for him to do; surely the dimensions of the earth are too big. He wonders about number 3. Given the fact that celestial objects change their orientation depending on your north-south location and that changes in your north-south location corresponds to turns (or angles) on a great circle around the earth, it certainly seems plausible to him that one can connect angle measurements on the earth to what is happening in the sky.[5]

A case in point: Celestial objects in the sky will look very different at the North Pole than they do at the equator and a change in location from the equator to the North Pole corresponds to a quarter turn measuring 90 degrees on the earth. The distance corresponding to a 90-degree turn on the earth is much too large to measure directly but he wonders if isn't possible to directly measure $\frac{1}{90}$th of that distance, namely, the distance on the earth that corresponds to 1 degree.

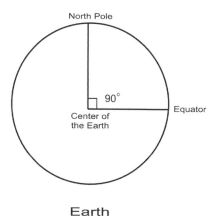

Earth

To get a handle on the problem, he decides to first find out if he can, by using objects in the sky and on the surface of the earth alone, determine that the equator and the North Pole are 90 degrees apart. He creates a model of a great circle of longitude on the earth and imagines two tall towers atop two buildings, one at the equator and the other at the North Pole (sizes not drawn to scale).

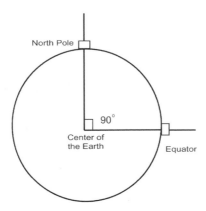

Since the sun is so far away, he knows that its rays can be considered to be essentially parallel by the time they reach the earth.

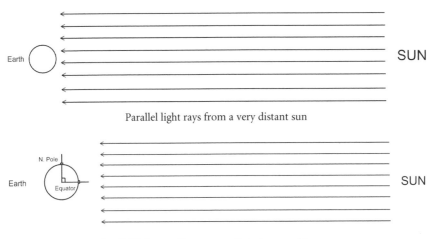

Parallel light rays from a very distant sun

Parallel light rays from a very distant sun and towers

Observe that the angle the sun's rays make with the tower on the equator is different from the angle they make with the tower at the North Pole. If the sun is directly overhead of the tower at the equator (as the diagram implies), then the angle its rays make with the structure is 0 degrees (they are parallel to each other). At the same time, its rays will make an angle of 90 degrees with the tower at the North Pole.

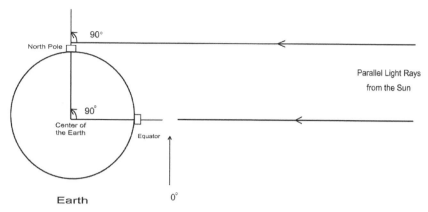

Angle between sun's rays and the towers at the Equator and North Pole
equals their angular separation on the Earth (90 degrees)

The angle that the sun makes with the tower at the North Pole matches perfectly with the actual angular separation on earth between the pole and the equator. In this case at least, if you can measure angles made by the sun's rays, then you can in principle use those measurements to determine the angular separation of these two buildings. Is this always the case?

Let's now look at the case where one of the buildings is not at the North Pole, but at a location halfway between the North Pole and the equator. This corresponds to an angular separation on the earth of 45 degrees.

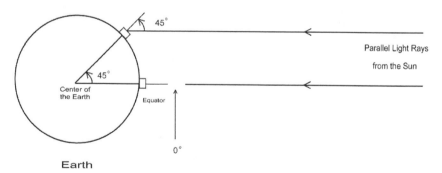

Angle between sun's rays and the towers equals their
angular separation on the Earth (45 degrees)

See that the angular separation between the locations on earth matches the angle between the sun's rays and the tower on the building 45 degrees away from the equator. It is a known fact from geometry that for two parallel lines such as the light rays in these two figures this correspondence will always hold. This general result is given here:

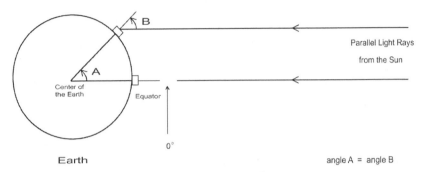

Arbitary Angles (A = B) with Earth included

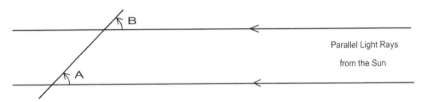

Arbitary Angles (A = B) without Earth

Although angles A and B are equal in value, a monumental difference is involved in obtaining their measures. Angle A resides deep inside the earth and is effectively inaccessible, whereas angle B lies on the surface and can be measured directly. And once more, the almost divine gift of substituting one set of objects or procedures by a more convenient set of objects or procedures presents itself to us—this time allowing the simpler task of measuring angle B to be substituted for the much more difficult one of measuring angle A—with no loss of information.

Although measuring angle B can be substituted for measuring angle A in theory, Eratosthenes still had to figure out a way to do it. How do you measure the angle that the sun's rays make with a tower? Since the sun doesn't discriminate in sharing its wealth, Eratosthenes knew that its rays make the same angle with any tower or object in a given location—meaning that he could use the angle the sun's rays make with a small stick in the place of a tower to obtain the angle (this can be done by using the shadow made by the stick).

He recalls reading about a curious situation in a town called Syene (present-day Aswan, Egypt), almost 490 miles directly south of Alexandria. On the first day of summer when the sun was at its highest point, it had been observed that its light shown directly down onto the water in a deep well (with its shimmering reflection clearly visible).

Eratosthenes realized that the only way that this could happen was if the sun was directly overhead Syene at this time. This would also mean that a tall tower or a thin stick held vertically in Syene would make virtually no shadow—implying that the angle between a tower and the sun's rays would equal 0 degrees as in the earlier scenarios.[6]

He decides to make an identical measurement using shadows created by the sun in Alexandria on the same day of the year at the same time (when the sun is at its highest point). He finds that the angle between a tower or stick here and the light rays is around 7 degrees. He now has the problem bagged.

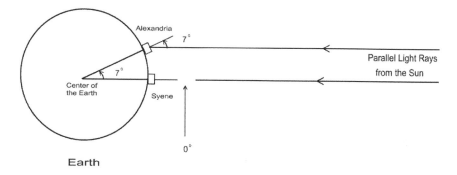

Since Syene is almost directly south of Alexandria, the two towns are very nearly on the same circle of longitude. This means that the angular separation between Syene and Alexandria must be close to 7 degrees. Given that their mileage separation is approximately 490 miles, Eratosthenes ecstatically concludes that on this big circle which goes around the earth, 7 degrees of turning must roughly equal 490 miles. Dividing by 7 then implies that 1 degree corresponds to approximately 70 miles.

Now he can apply Amar's method number 3 and multiply 360×70 miles to obtain 25,200 miles as the distance around the earth. This answer is off by only a few percentage points of the answer we know to be true today.[7]

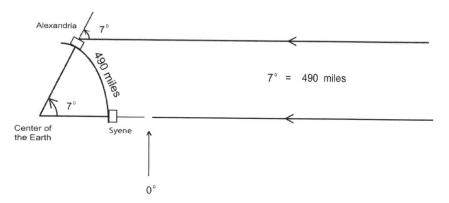

Making a Spectacular Statement

Amar's humble studies on circles, done in the dirt generations before, have now made a spectacular statement! The results obtained by him and others, in the hands of Eratosthenes, have forced the earth to give up one of her greatest secrets—the mystery of her size. Eratosthenes was able to do this not by the direct method of going around the earth and measuring its length, but rather by staying at home and simply creating models, using a few measurements, and manipulating symbols.

Awareness of the fact that one thing in nature can be substituted for another in a crushingly advantageous way is what enabled Eratosthenes to do this. The key point being that while we as humans do discriminate between forms, a small circle in the dirt as being something drastically different from a large one that goes around the earth, many of the fundamental rules that we learn about them do not. A circle is a circle is a circle and once the essential laws have been obtained wherever that is, they apply across the spectrum of forms, from a small circle drawn in the dirt more than 4,000 years ago to a modern circle as large as the very earth itself. It simply makes no difference.

We say it again: For people the two circles are vastly different, the circle drawn in the dirt can be taken in with a glance while circuits around the earth are so big that as far as we know it would take more than 1,700 years, from the time of Eratosthenes, before someone was capable of actually traversing one. To make a few measurements using a small stick and a few symbolic calculations and then be able to connect them to the distance around an enormous planet, without actually going around that planet, is something very, very special indeed.

The realization that mathematics can be used to spectacularly access "the inaccessible and inconvenient" is certainly one of the main reasons why human beings have cared about it for so long. It is certainly a major reason why scientists often wax eloquent when describing the subject. Applying math to solve problems is one of the most deeply satisfying and worthwhile activities we collectively do.

Charles Darwin is often quoted as saying: "mathematics seems to endow one with a new sense," and while mathematics certainly does not fit most scientific definitions of what a physiological sense is, it does give us a huge lens to peer into a world that we cannot directly perceive, revealing patterns and connections which can be leveraged to "magnify" our understanding of the world.[8]

This occurs because just as we can translate our daily happenings and thoughts into sounds (speech) or visible marks (writing) with great advantage, so too can we translate other meaningful types of information into mathematical diagrams and symbols to great advantage. With the exception of language whose utter centrality to human affairs there is no denying, the ability to advantageously translate or convert one set of objects or procedures into another more convenient set of objects or procedures roars the loudest in mathematics.

Conclusion

As the stated calculation which Eratosthenes used to estimate the distance around the earth, "360 × 70 = 25,200" plays a significant role in this chapter. What is curious to note, however, is that this same calculation can also play extremely minor roles as well. If 360 assorted pieces are in each of 70 boxes of cereal, then "360 × 70 = 25,200" also gives the total number of pieces of cereal in all of the boxes.

The same calculation describing something as significant as a valid estimate of the distance around our planet, on the one hand, can also describe something as mundane as the total number of pieces of cereal in a collection of seventy boxes, on the other. There are infinitely many other potential situations, equally mundane or significant, that can also be described by this calculation (of course, both "mundaneness" and "significance" are in the eye of the beholder).

The existence of such situations means that simple mathematical statements such as "360 × 70 = 25,200" have much in common with language words and language statements. For example, the single word "house" can simultaneously be used as a significant metaphor in a presidential address to the nation or as a commonplace noun in ordinary coffee shop banter.

The word "house" can also represent millions of different existing structures, and ultimately millions of different structures yet to be built. These structures vary in size, inhabitants, color, shape, location, value, composition, and so on. In spite of this, they all still have enough in common to be represented by the same basic five-letter word.

In the case of the mathematical statement, "360 × 70 = 25,200" the different structures are replaced by the multitude of different and diverse situations. Some of the situations described by this expression can be immensely significant while others can be commonplace—even boring.

This is an important property to realize about all symbolic mathematical expressions. They can describe literally an infinite cloud of potential situations, making them extremely general just like language words and sentences, perhaps even more so. Thus, in learning how to manipulate expressions and diagrams we are in a sense learning how to manipulate the entire cloud of possibilities directly or potentially described by that expression or diagram.[9] This has happened over and over again in this chapter with every mathematical thing we did (e.g., methods for finding the distance around circles in Amar's backyard, which will work in principle on any circle in the universe, the equality of corresponding angles which will work for trillions of angles and trillions of towers on trillions of different planets).

The conceptual significance of this chapter becomes clear now and is this: Through the judicious employment of symbols, diagrams, and calculations, mathematics enables us to acquire significant facts about extremely significant things (universal laws, even), not by first forging out into the cosmos with teams of scientists, but rather from the comforts and confines of coffee tables in our living rooms! This was certainly the case with both Amar and Eratosthenes.

But while true in our tale here (and in thousands of others), it is important to note that not every random and local investigation will produce truths that are both eternal and significant. The key to making it work is perhaps best encapsulated in the words attributed to the eminent mathematician David Hilbert: "The art of doing mathematics consists in finding that "special case" which contains all the germs of generality."[10] Local investigations of circles, angles, parallel lines, and numbers all have the qualities of that "special case."

The stage is now set to transition to the next phase of the book—a detailed conceptual study of the four fundamental operations of addition, subtraction, multiplication, and division. These operations form an important core in grade school mathematics curricula around the world. We have already seen that even something as elementary as basic numeration is no trivial matter, and we will find the same to be true of these four classical operations as well.

Throughout what follows, the often-mentioned computational efficiency of the HA notation will be on full display. However, there is also much "conceptual manna" to be gleaned from the coin numeral and abacus models that we have developed. So instead of tossing these systems aside as interesting curiosities and forgetting about them, we will now recalibrate them to become potent weapons of exposition. These models then will comprise a crucial portion of the conceptual arsenal which we will employ full force in our efforts to give readers a newfound appreciation for the elementary arithmetic already in their possession.

II

THE SPECTACULAR FUSION OF CALCULATION WITH WRITING

The process of "pen reckoning," as calculation with Hindu-Arabic numerals was sometimes called did not efface itself as it occurred, and thus could be easily checked; and calculating and recording could be done with the same symbols.

—Alfred W. Crosby, contemporary historian,
writer, author of *The Measure of Reality:
Quantification and Western Society, 1250–1600*[1]

Author's Note

THE PHYSICIST, IN EXPLAINING projectile motion to the novice, as a first step, ignores air resistance (or friction) because it makes the analysis easier for beginners to digest—focusing on some of the crucial fundamentals without overly complicating things.

I fully concur with this plan of attack as well as with the thoughts of Renaissance educator Wolfgang Ratke when he states: "Before the learner has a notion of the thing itself, it is folly to worry him about its accidents."[1]

In this text, I have taken a similar tack and have, for the purposes of exposition, chosen to view the elementary operations of arithmetic as they look through the lens of whole numbers only, ignoring the details of how they look when negatives, fractions, and irrational numbers enter the arena. I make no apology for this approach.

For readers who would like to see more examples of some of the conceptual procedures involved in the next five chapters please visit: www.howmathworks.com.

5

Numeral Formations
Come to Arithmetic

1234567890
2345678901
3456789012
4567890123
5678901234

"Should you just learn the rules and not the concepts or should you only understand the concepts and not the rules?" is like saying "Would you rather be deaf or blind?" If you are blind, you do not see, if you are deaf, you do not hear. On the whole we prefer to have both faculties.

—Sir Michael Atiyah (paraphrased), British mathematician,
Fields Medal and Abel Prize laureate[1]

"MAKE SURE YOU SHIFT the 252 in the second row," the teacher exclaimed! The startled child, for the third straight time in front of the classroom, had calculated the product of 42 × 67 in the following manner,

$$\begin{array}{r} \overset{1}{4}2 \\ \times\ 67 \\ \hline 294 \\ 252 \\ \hline 546 \end{array}$$ as opposed to $$\begin{array}{r} \overset{1}{4}2 \\ \times\ 67 \\ \hline 294 \\ 2520 \\ \hline 2814 \end{array}$$; thus making a mistake that has undoubtedly been

committed by millions of young students since the 1500s. With the introduction of the HA system into arithmetic (and its practice of using the same symbols in different locations), keeping the numerals in the right formation had become a serious affair.

But numeral formations are only a piece of the puzzle. To perform calculations correctly, one must also learn the basic facts of how the ten fundamental numerals {0, 1, 2, 3, 4, 5, 6, 7, 8, 9} interact with each other in regards to the elementary operations of addition, subtraction, multiplication, and division. Correctly apply these basic facts while keeping the numerals in the right formations and you get the right answer. Apply the basic facts improperly (saying 5 + 4 is 8 instead of 9, for instance) or put the numerals in incorrect formations (thus making what should be 2520 into 252) and you don't. It is as simple as that.

If, as is often stated, the introduction of the HA system into arithmetic was revolutionary, then it is no stretch to say that the need to memorize these basic facts along with placing numerals in the right formations are two very important pieces in this revolution. Hundreds of basic facts have to be learned to successfully add, subtract, multiply, and divide just the counting numbers in this system. If we add the rules for putting the numerals into the right formations, as well as the techniques for applying these numerals to solve various problems—not to mention fractions and negative numbers—then the number of concepts that must be learned in elementary arithmetic mushrooms to many more. This can be intimidating to someone who has yet to master the system.

The need to learn so many basic facts effectively means that the introduction of the HA system has turned out to be a double-edged sword. With one edge being a blessing because it allows for the historically revolutionary idea that arithmetic can in principle be learned by anyone. The average person nowadays is fully capable of using elementary arithmetic in ways that even people who were good at mathematics would have found time-consuming and difficult to perform in other systems—unaided by a mechanical device such as an abacus. The opposite edge being a curse because the need to learn so many facts can at times seem overwhelming. In addition, the very efficiency of this system may easily prevent a deep understanding of what is really happening with calculation in elementary arithmetic; thus masking the absolute miracle in writing that has been achieved by the positional representation of numbers.

Our goal in this chapter as well as in subsequent ones will be to expound upon how the need for learning numeral formations and so many basic facts arose, as well as the benefits gained by fulfilling this need. Later in the book, we will discuss some of the pitfalls that have arisen in education because of all of this. We begin our discussion here by revisiting addition once again.

Addition

In what follows, we will freely alternate between the three systems discussed so far: the HA numerals, the abacus rods, and the coin numerals. Although their origins lie in the tangible physical device, our abacus rods have now become entities in their own right. In what follows, they will be used as independent objects in ways that would be difficult to mimic with an actual abacus. Working with the abacus rods and coins will give us conceptual information on how to proceed in constructing a recipe or algorithm for addition in terms of the HA script. We begin by observing how adding 24 and 43 plays out on abaci rods:

We can proceed horizontally, but it is more convenient to arrange the columns vertically so that their denominational values fall more easily into alignment. This rearrangement will become decidedly advantageous when carries are involved. Beginning with the ones column (or denomination) and moving to the left, we have:

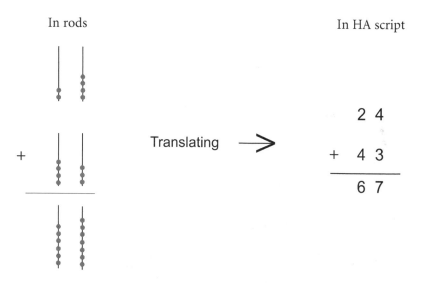

In rods   In HA script

Translating $\longrightarrow$

$$
\begin{array}{r}
2\ 4 \\
+\ \ 4\ 3 \\
\hline
6\ 7
\end{array}
$$

The next example of 56 + 25 = 81 demonstrates how we handle situations when more than ten beads end up on one of the rods:

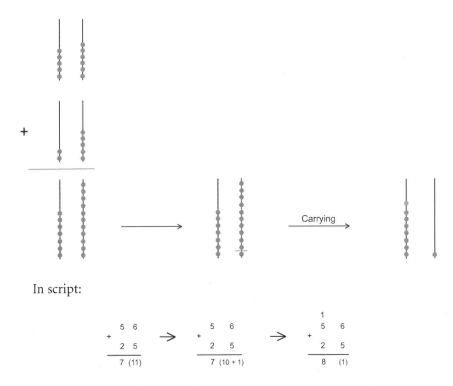

In script:

									1	
	5	6	→		5	6	→		5	6
+				+				+		
	2	5			2	5			2	5
	7	(11)			7	(10 + 1)			8	(1)

In these examples, the rods are doing all of the work for us. This is nothing new since the traditional use of the abacus has been as an aid in calculation for a written script. This is similar to the role coin numerals played in helping us obtain the answer to the addition of fifteen and twelve in chapter 3. Unlike the situation with coin numerals, however, the HA script is a direct rendering of the abacus rods; so even if a person did not know how to conceptually add in script they could still write the correct answers using only a knowledge of how adding the "ten original designs" on the abacus rods translates to HA script. This means that as our skill with the rods improves, this should directly translate to our skill using the HA numerals. Our skill with the rods will improve dramatically if we simply bite the bullet and memorize what the results will be, when we pair-wise add all of the "ten original designs" on the abacus rods. These results are listed here (the tables read as A + B):

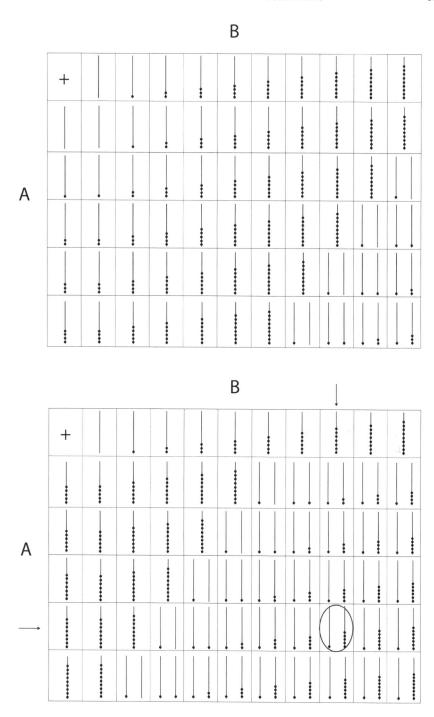

Abacus Rod Addition Table

Using the table is pretty straightforward as the example of adding a rod with eight beads to one with seven beads shows. We look for the eight rod in the A column and the seven rod in the B row. The arrows identify these. The circled entry indicates their sum after the carry. We conclude that:

In HA script, this directly translates to: $8 + 7 = 15$.

All of the results in the abacus rod addition table can be directly translated into the more compact HA script. These results yield the following addition table in script:

+	0	1	2	3	4	5	6	7	8	9
0	0	1	2	3	4	5	6	7	8	9
1	1	2	3	4	5	6	7	8	9	10
2	2	3	4	5	6	7	8	9	10	11
3	3	4	5	6	7	8	9	10	11	12
4	4	5	6	7	8	9	10	11	12	13
5	5	6	7	8	9	10	11	12	13	14
6	6	7	8	9	10	11	12	13	14	15
7	7	8	9	10	11	12	13	14	15	16
8	8	9	10	11	12	13	14	15	16	17
9	9	10	11	12	13	14	15	16	17	18

Hindu-Arabic Addition Table

We will sometimes abbreviate the HA addition table as simply the addition table. Notice that these two tables appear to say something only about the ones and tens rods (or ones and tens place) but, as before, the rules so discovered on a particular column or place value are rules that translate to any column or place value. The following demonstration shows how to apply the results in the addition tables to the ones, tens, and hundreds places:

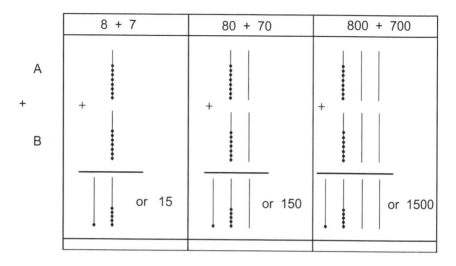

8 + 7	80 + 70	800 + 700

These examples illustrate that calculations of 8 + 7, 80 + 70, and 800 + 700 amount to playing the same game in different locations:

For 8 + 7, the game is played in the ones and tens places.
For 80 + 70, the same game is played in the tens and hundreds places.
For 800 + 700, the game is played in the hundreds and thousands places.
For 8,000 + 7,000, not shown, the game is played in the thousands and ten thousands places.

This extends to higher places as well. The essential rule in every case is that 8 + 7 = 15 and this is what is codified in the addition table. Knowing this rule alone is not enough, however. To successfully engage it we must learn how to properly handle the fact at whatever position it occurs in a numeral formation—this is a distinctly different task.

We can shorten the addition procedure if we deal with the carries up front—as we add, rather than waiting until after the addition is completed. In the addition of 56 and 25 we demonstrate this by moving the carries into the open space above the rods:

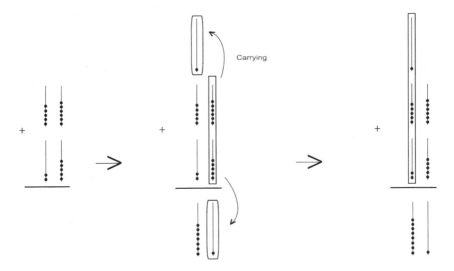

In script:

The next demonstration adding 678 and 545 shows how this shortened algorithm looks when multiple carries are involved:

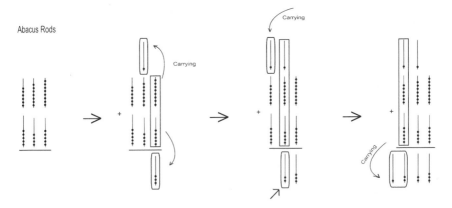

In script:

Here we have done the primary additions, $8 + 5 = 13$, $7 + 4 = 11$, and $6 + 5 = 11$ plus the modifications necessary to handle the additional "*1s*" in the carry. These primary additions can all be found in the HA addition table. This sum can also be done with coins as well. The answer is 1223 in all three systems.

The procedures involving HA script and the abacus rods can both be done on a single diagram. Doing this with the HA yields:

$$
\begin{array}{r}
1\ 1 \\
6\ 7\ 8 \\
+\quad 5\ 4\ 5 \\
\hline
1\ 2\ 2\ 3
\end{array}
$$

This is clearly the most compact and efficient of the three methods. Consequently, it will be to our great advantage to master this method, and an important cog in making it work is to memorize outright the HA addition table. This table will serve as our "alphabet" for addition. Once this is accomplished, we can efficiently use the nice vertical procedure for the HA numerals to add any two whole numbers. We demonstrate this for the following addition:

$$
\begin{array}{r}
111\ 1\ 1\ 1\ 1\ 1 \\
2\ 5\ 7\ 6\ 0\ 0\ 8\ 0\ 9\ 6\ 7 \\
+\quad 1\ 6\ 0\ 8\ 8\ 9\ 5\ 9\ 8\ 7\ 4 \\
\hline
4\ 1\ 8\ 4\ 9\ 0\ 4\ 0\ 8\ 4\ 1
\end{array}
$$

Primary additions used from the addition table this time are:

$$7 + 4, 6 + 7, 9 + 8, 0 + 9, 8 + 5, 0 + 9, 0 + 8, 6 + 8, 7 + 0, 5 + 6, 2 + 1$$

It is certainly possible to do this calculation in coin numerals or abacus rods but the amount of work as well as the space used will be much greater for both.

If we only add two numbers, the most we can carry to another column is a 1. If, however, we add three or more numbers at once, it is possible that values larger than 1 may be required in a carry. In such cases, memorization can be combined with mental calculation to obtain the result. Try this with 687 + 796 + 587.

If we committed the addition table to memory and practiced representing numbers using HA numerals only, we could dispense with the rods and beads entirely, essentially creating a new numerical language not only for representing numbers but also for adding them too. Future generations learning only this new numerical language might not even be aware of the rods and beads from which they were derived.[2] This approximates what happened and is not wholly unlike what happens when new technologies replace old ones.

The reader should keep in mind that our conversations involving HA numerals are broad and widespread. As discussed earlier, mathematical statements (whether in the form of tally marks, coin configurations, abacus rods, or HA numerals) all act like language words and sentences which means that they can be used to describe an enormous number of possibilities. Some of those possibilities in the case of "4589 + 2718 = 7307" include: 4589 people + 2718 people = 7307 people, $4589 + $2718 = $7307, 4589 books + 2718 books = 7307 books, and so on.

A key component that distinguishes the HA numerals from the others is that, like language, the script requires a heavy investment in memorization. The scheme involving coin numerals does not require the memorization of a table for addition—exactly what to combine and how to change denominations is very apparent. However, the gains acquired in this trade-off are more than worth the trouble. Not only is it now possible to make the representation of numbers alphabetic in HA script, but with the memorization of the one hundred entries in the addition table, it becomes possible to alphabetize the operation of addition itself![3]

The trillions upon trillions of different additions between any two numbers no matter their size can be reduced to the one hundred additions in the addition table. Those one hundred additions act as a closed alphabet for all additions of two numbers.

Subtraction

Subtraction Using Coin Configurations

What about the opposite process of taking objects away from a collection as opposed to adding them? This operation is called subtraction. Now that we have seen that it is possible to alphabetize addition, the immediate question becomes can we do it with subtraction too? Can lightning strike twice in the same way? We will find that indeed it can but this time with a twist.

Let's first look at subtraction with coin numerals and observe how 784 − 362 plays out:

In script this reads as 784 − 362 = 422.

Sometimes it is not so easy, and before we can simply cross out coins in the subtraction, the coins have to be first rearranged to make it work. The example of 654 − 467 demonstrates this:

Here we can easily take out the four (H) coins in four hundred from the six (H)s above it in six hundred fifty-four, but as it now looks, there are not enough (T) and (I) coins, respectively, in 654 for our procedure of striking out to work in its present form. This turns out only to be a mild discomfort, for we can convert a (H) into ten (T)s and a (T) into ten (I)s. Doing this and then completing the subtraction yields:

Translating this back to script allows us to conclude that 654 − 467 = 187. Changing a higher denomination coin into ten of a smaller denomination is a frequent occurrence in subtraction and corresponds to what is traditionally called borrowing.

The notion of taking away from a collection (subtraction) is the opposite of adjoining to a collection (addition). This inverse relationship is reflected in our coin manipulations (calculations): Where in the case of addition, ten coins of a lower denomination are often grouped together to yield a coin of a higher denomination, while conversely in the case of subtraction, a coin of a higher denomination is often ungrouped to yield ten coins of a lower denomination.

As long as we are subtracting a smaller whole number from a larger whole number, borrowing ten coins from a higher denomination will always supply us with enough coins to complete the subtraction: The reason is that the most number of coins we can have for any given number (assuming the grouping conventions have been obeyed) is nine and when we decompose a higher denomination into a lower one we get ten new coins.[4]

Translating Subtraction to HA Script

This method of using coins will work for any subtraction of a smaller whole number from a larger one. This gives us a general method for subtraction using coin numerals that is both systematic and reproducible (meaning that anyone who uses it correctly will always arrive at the same answer). Unfortunately, just like in addition with coins, the method is very "wordy." Our next steps are to see how the general method discussed earlier looks using the abacus rods and to then compress these results into a method or algorithm involving only HA script. We first look at 784 − 362:

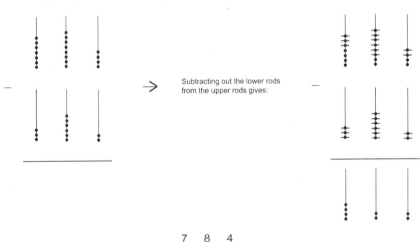

Subtracting out the lower rods from the upper rods gives:

In script this translates as:

$$
\begin{array}{r}
7\ 8\ 4 \\
-\ 3\ 6\ 2 \\
\hline
4\ 2\ 2
\end{array}
$$

How do situations involving a borrow look with rods? We revisit 654 – 467. As we saw with coins, there are not enough beads in the ones column of the 654 to cancel out the 7 beads in the ones column of the 467. To handle this, we must convert and align:

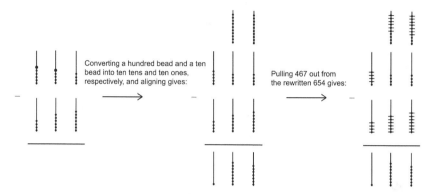

In HA script this becomes:

$$
\begin{array}{r}
6\ 5\ 4 \\
-\ \ 4\ 6\ 7 \\
\hline
\end{array}
\quad
\xrightarrow[\text{\& the 6 hundreds gives:}]{\text{Borrowing from the 5 tens}}
\quad
\begin{array}{r}
\ \ \ \ 10\ 10 \\
5\ \ 4\ \ 4 \\
-\ \ 4\ \ 6\ \ 7 \\
\hline
\end{array}
\quad
\xrightarrow[\text{\& 14 tens respectively gives:}]{\substack{\text{Adding 10 and 4 in both}\\ \text{cases to obtain 14 ones}}}
\quad
\begin{array}{r}
5\ \ 14\ \ 14 \\
-\ \ 4\ \ \ 6\ \ \ 7 \\
\hline
1\ \ \ 8\ \ \ 7
\end{array}
$$

Thus we have that 654 – 467 = 187.

As with addition, the rods are presently doing the work for us. To make a practical scheme or algorithm for subtraction in HA script, it will help if we memorize all of the possible situations we can encounter when subtracting two digits within the system.

Subtraction differs from addition, however, in that borrowing or decomposing is involved as opposed to carrying. What this means is that when a borrow is involved we must change the form of the number before a fundamental subtraction can take place; this never happens in addition. In adding, we perform the fundamental additions from the table first and then the carries. We will clarify the distinction with the examples of 305 + 998 and 939 – 587:

$$
\begin{array}{r}
1\ 1\ \ \ \\
3\ 0\ 5 \\
+\ \ \ 9\ 9\ 8 \\
\hline
1\ 3\ 0\ 3
\end{array}
$$

Fundamental additions used from the addition table are: 5 + 8 = 13, 0 + 9 = 9, and 3 + 9 = 12.

	9	3	9
-	5	8	7

Borrowing from the
9 hundreds gives:

$\rightarrow$

		10	
	8	3	9
-	5	8	7

Adding 10 and 3 in the
tens column to obtain 13
tens gives:

$\rightarrow$

	8	13	9
-	5	8	7
	3	5	2

Fundamental subtractions involved are: 9 − 7 = 2, 13 − 8 = 5, and 8 − 5 = 3.

A key difference in the two calculations in the example is the 13 − 8 that occurs in the middle subtraction after a borrow. Thirteen is two numerals long, not one. This is true in general for fundamental subtractions which require a borrow, meaning that we have to also deal with two-digit long strings. This never occurs with addition since the carry happens after each of the numerals have been added in a given column.

Another key difference is that what we see in the initial subtraction diagram on the left is not what ends up being subtracted. In the final diagram, we end up performing the fundamental subtractions 13 − 8 and 8 − 5, neither of which appears in the original diagram of

	9	3	9
-	5	8	7

Only the fundamental subtraction 9 − 7 appears in both the original and final subtraction diagrams.

All of this means that in HA numerals we effectively have two types of subtractions: those that require a borrow and those that don't. Unlike addition, the fundamental HA table for subtraction splits into two: subtractions between digits requiring a borrow and subtractions between digits not requiring a borrow. We list both situations here in a yes/no format:

Yes/No Borrowing[a]

B

−	9	8	7	6	5	4	3	2	1	0
9	N[b]	N	N	N	N	N	N	N	N	N
8	Y	N	N	N	N	N	N	N	N	N
7	Y	Y	N	N	N	N	N	N	N	N
6	Y	Y	Y	N	N	N	N	N	N	N
5	Y	Y	Y	Y	N	N	N	N	N	N
4	Y	Y	Y	Y	Y	N	N	N	N	N
3	Y	Y	Y	Y	Y	Y	N	N	N	N
2	Y	Y	Y	Y	Y	Y	Y	N	N	N
1	Y	Y	Y	Y	Y	Y	Y	Y	N	N
0	Y	Y	Y	Y	Y	Y	Y	Y	Y	N

A (row labels)

[a]Table reads as A − B
[b]Y ≡ Need to Borrow or Decompose and N ≡ No Need to Borrow or Decompose

This leads to the following two tables based on the above yes/no results:

Subtraction Table (No Borrowing Required)

B

−	9	8	7	6	5	4	3	2	1	0
9	0	1	2	3	4	5	6	7	8	9
8		0	1	2	3	4	5	6	7	8
7			0	1	2	3	4	5	6	7
6				0	1	2	3	4	5	6
5					0	1	2	3	4	5
4						0	1	2	3	4
3							0	1	2	3
2								0	1	2
1									0	1
0										0

A (row labels)

Subtraction Table (Borrowing Required)

B

.	9	8	7	6	5	4	3	2	1	0
10 + 8 = 18	9									
10 + 7 = 17	8	9								
10 + 6 = 16	7	8	9							
10 + 5 = 15	6	7	8	9						
10 + 4 = 14	5	6	7	8	9					
10 + 3 = 13	4	5	6	7	8	9				
10 + 2 = 12	3	4	5	6	7	8	9			
10 + 1 = 11	2	3	4	5	6	7	8	9		
10 + 0 = 10	1	2	3	4	5	6	7	8	9	

(A labels the left column of the table.)

These two tables will serve as an "alphabet" for subtraction. The twist, or what makes the procedure different from what we did in addition, is that the fundamental situations split into two tables as opposed to one. Using only the tables now for the subtraction 728 – 483, without the aid of abacus rods, gives:

$$
\begin{array}{r} 7\ 2\ 8 \\ -\ 4\ 8\ 3 \\ \hline \end{array}
\xrightarrow[\text{table gives:}]{\text{No borrowing}}
\begin{array}{r} 7\ 2\ 8 \\ -\ 4\ 8\ 3 \\ \hline 5 \end{array}
\xrightarrow[\text{borrowing table gives:}]{\text{Borrowing and using}}
\begin{array}{r} {}^{10}\!6\ 2\ 8 \\ -\ 4\ 8\ 3 \\ \hline 4\ 5 \end{array}
\xrightarrow[\text{table gives:}]{\text{No borrowing}}
\begin{array}{r} {}^{10}\!6\ 2\ 8 \\ -\ 4\ 8\ 3 \\ \hline 2\ 4\ 5 \end{array}
$$

Fundamental subtractions:
Not requiring a borrow: 8 – 3 = 5; 6 – 4 = 2
Requiring a borrow: 12 – 8 = 4

Shortening the Subtraction Algorithm

Our next goal is to shorten the subtraction algorithm when borrowing is involved. In the first subtraction (784 – 362), where no borrowing was required, we can essentially do the subtraction in a single step just as we did in addition. However, for the subtraction 728 − 483, which involved borrowing, the work required several steps. The biggest problem involved the fact that we had to make changes to the top number to complete the subtraction; that is, although we started with a subtraction of 7 − 4 in the hundreds column, we ended up actually subtracting 6 − 4 due to the borrow.

This means that if we shorten the algorithm to a single diagram we have to do something to indicate the changes that can occur to a digit when we borrow. In performing addition, the digits of the top number in the original addition stay the same, and when a carry is needed, the changes happen in the space above the number.

However, in subtraction the changes can occur in the digits of the top number, itself, as well as in the space above. To represent such subtractions in a single step, we must identify what we are doing by tagging the digits that are changing. We illustrate one method by reconsidering 728 − 483:

We simply combine all of the steps in the previous example into a single one, crossing out the 7 to indicate that it is being replaced by a 6:

$$
\begin{array}{cc}
\begin{array}{r}
7\ \ 2\ \ 8 \\
-\ \ 4\ \ 8\ \ 3 \\
\hline
\end{array}
&
\begin{array}{r}
\overset{6\ \ 10}{\cancel{7}\ \ 2\ \ 8} \\
-\ \ 4\ \ 8\ \ 3 \\
\hline
2\ \ 4\ \ 5
\end{array}
\end{array}
$$

Using the subtraciton tables and borrowing gives:

The fundamental subtractions from the table remain the same.

The subtraction of 64,053 − 28,236 shows how messy a subtraction problem can be when multiple borrows are required:

$$
\begin{array}{cc}
\begin{array}{r}
6\ \ 4\ \ 0\ \ 5\ \ 3 \\
-\ \ 2\ \ 8\ \ 2\ \ 3\ \ 6 \\
\hline
\end{array}
&
\begin{array}{r}
\overset{\ \ \ \ \ \ \ \ \ \ \ \ \ \ \ \ 10}{\overset{5\ \ \ 3\ \ 10\ \ 4\ \ 10}{\cancel{6}\ \ \cancel{4}\ \ 0\ \ \cancel{5}\ \ 3}} \\
-\ \ 2\ \ 8\ \ 2\ \ 3\ \ 6 \\
\hline
3\ \ 5\ \ 8\ \ 1\ \ 7
\end{array}
\end{array}
$$

Using the subtraciton tables and borrowing gives:

Fundamental subtractions from the table:
Not requiring a borrow: 4 − 3 = 1; 5 − 2 = 3
Requiring a borrow: 13 − 6 = 7; 10 − 2 = 8; 13 − 8 = 5

The methods demonstrated here, in HA numerals, are all of a type called the decomposition method. This method of subtraction evidently caught on like wildfire in the United States after it was reintroduced in the late 1930s by the great educator William A. Brownell and has become the predominant way that subtraction is now taught in America. Before this time, two other methods called the equal additions method and the Austrian method, respectively, were used nearly as much. The Austrian method remains popular in Europe. The latter two methods require more memorization to use than the decomposition method but work just as well and don't vandalize the vertical diagram as much.[5] All three methods allow us to perform subtractions within a single expression.

A variant of the decomposition method is also still quite popular in the United States today. This approach abbreviates the procedures just a bit further. We demonstrate this for subtracting 728 − 483:

$$
\begin{array}{ccc}
 & & \overset{6}{} \ \overset{1}{} \\
7 \ 2 \ 8 & & \overset{6}{\cancel{7}} \ 2 \ 8 \\
\underline{- \ \ 4 \ 8 \ 3} & \xrightarrow[\text{and borrowing gives:}]{\text{Using the subtraciton tables}} & \underline{- \ \ 4 \ 8 \ 3} \\
 & & 2 \ 4 \ 5
\end{array}
$$

The only difference between the two methods shown for calculating 728 − 483 is that the $\overset{10}{{}_2}$ in the earlier subtraction is shortened to $\overset{1}{{}_2}$ in the abbreviated method. We read the $\overset{1}{{}_2}$ as 12. Otherwise, the content in both methods is the same.

The subtraction 64053 − 28236 abbreviates as follows:

$$
\begin{array}{cccccc}
 & & & & \overset{1}{} & \\
 & & 5 & 3 & 1 & 4 & 1 \\
6 \ 4 \ 0 \ 5 \ 3 & & \cancel{6} \ \cancel{4} \ 0 \ \cancel{5} \ 3 \\
\underline{- \ 2 \ 8 \ 2 \ 3 \ 6} & \xrightarrow[\text{and borrowing gives:}]{\text{Using the subtraciton tables}} & \underline{- \ 2 \ 8 \ 2 \ 3 \ 6} \\
 & & 3 \ 5 \ 8 \ 1 \ 7
\end{array}
$$

Here the $\overset{10}{{}_3}$ in the earlier subtraction has twice been replaced by $\overset{1}{{}_3}$ which is interpreted as 13 and the $\overset{10}{{}_0}$ has been replaced by $\overset{1}{{}_0}$ which is interpreted as 10.

This abbreviated method introduces an inconsistency between how we add and subtract. For instance, in our earlier addition of 305 + 998, we read $\overset{1}{{}_3}$ and $\overset{1}{{}_0}$ as 1 + 3 = 4 and 1 + 0 = 1, respectively, while we read them in the abbreviated subtraction 64,053 − 28,236 as 10 + 3 = 13 and 10 + 0 = 10, respectively. This inconsistent treatment, in addition and subtraction, of arrangements that look the same is undoubtedly a source of difficulty for those learning the abbreviated method, particularly if the distinction is not pointed out. It is not unlike the irregular treatment in English pronunciation of the sequence of letters "cycle" in the words "re*cycle*" and "bi*cycle*" or of the sequence of letters "ough" in the words "t*ough*" and "th*ough*."

Regardless of the method used, subtraction succumbs to being "alphabetized" in HA numerals. What this means is that, as in addition, all subtractions (of the type "a larger whole number" minus "a smaller whole number") can be reduced by algorithm to the 100 subtractions coming from the two fundamental tables shown earlier. It is not enough, however, to simply memorize the information in the tables; one must also acquire the skill to handle the vertical methods that utilize this information. In the case of subtraction this can take quite some time. Once mastered, however, the subtraction techniques in HA notation are clearly superior in terms of convenience and compactness to the techniques we have used for the abacus rods and the coin numerals.

A Game of Checkers

A central theme of this book is the fact that one method can be spectacularly more advantageous to use than another method. Taken at face value, this obvious truth is not earth-shattering. What is valuable is that interpreting it in the context of mathematics can be conceptually illuminating in many ways.

Consider the game of checkers. We can play it with red and black checkers, nickels and dimes, bottles and cans, Fords and Chevys, even horses and elephants. Each set of pieces represents a form in which we can play the game; however, it is clearly more convenient to play the first two versions of the game as opposed to the latter two. So even though all piece sets are equivalent in that they offer different versions of the same game, they are not all equivalent in terms of user-friendliness.

Imagine, if you will, that humankind's first encounter with checkers had been the version where horses and elephants were used as game pieces. Playing this version would require great resources and wealth to play (lots of land, many horses and elephants, etc.) and consequently would in general be unavailable to the average citizen.

Now imagine that someone eventually discovers that it is immaterial whether the game pieces are alive or not, and decides to substitute small toy model horses and elephants in place of the real ones. This would allow the game to be played on a much smaller plot of land.

Our pioneer may go even further and realize that the pieces don't even have to look like horses and elephants (they just have to be two sets of distinctly different pieces) and decide to use a "coded set" of small red and black checkers on a small transportable red and black board. Implementation of this smaller more user-friendly version would have several immediate benefits:

- The game would be much more convenient and economical to play plus transportable, making it accessible to a much larger number of people, perhaps the entire citizenry.
- It would allow for systematic and conceptual study of the game to occur with far greater ease (the entire sweep of the game can now be seen all at once).
- Wider accessibility increases the chances of a truly great talent being exposed to the game.

Eventually this version or some other small version of the big game would dominate and most likely, over a period of time, the original horses and elephants version would be placed out to pasture. In the future, the only people that might even be aware of its existence would be those interested in the history of the game. Here, the substitution of the more advantageous form in place of the other offers such decisive advantages that it becomes historic for the game (and also the world, if the game is of great enough significance) and the replacement becomes permanent.

This gives an analogy of what often occurs with technological change: a technological discovery, in this case, learning to play checkers with small red and black checkers yields crucial advantages over the old way. These advantages turn out to be so great that the new way completely unseats the old way. This checkers scenario is obviously fictional but it metaphorically reflects what actually can and does occur in the symbolic world of mathematics. In elementary arithmetic, the old methods of calculating gave way to more efficient and user-friendly procedures. These newer techniques performed the task so much better that the switch was decisive, and over time knowledge of the earlier, comparatively more cumbersome (but still on occasion conceptually useful) methods have receded into the backdrop of history.

Conclusion

The natural problem of finding out the size of a collection after objects have been added to it or taken away from it has now been completely rerouted to the task of memorizing addition or subtraction tables and learning how to use the accompanying algorithm (i.e., the processes of placing the numerals into the proper formations and rearranging them to accommodate the carries and borrows). This is a feat of symbolic engineering.

Much repetition or drill is needed to learn both the tables and the algorithms, but once accomplished, the ease by which addition and subtraction, even of large numbers, can be performed makes the effort worthwhile. The algorithms are so effective, in fact, that it now becomes possible to use them without thinking about what is happening conceptually—this is especially true in the case of the more difficult subtraction algorithm. Not thinking about what is theoretically happening is quite fine if one has already conceptually mastered what is going on, but not so good if one does not understand at all conceptually why the algorithms work. Hopefully this chapter has helped shed some conceptual light on why the procedures for these two operations succeed in calculation.

The advantages gained by using the HA notation, while most impressive here in addition and subtraction, become truly monumental in the case of the higher order operations of multiplication and division. This is undoubtedly a major reason why the script ascended to the worldwide phenomenon it is today. We now turn our attention to these higher order operations.

6

The Symmetry of Repetition

The whole object of travel is not to set foot on foreign land; it is at last to set foot on one's own country as a foreign land.

—Gilbert Keith Chesterson, British author,
literary and social critic[1]

WE EAT THREE MEALS A DAY, go to work five days out of seven, and grocery shop perhaps once a week—repetitive acts are a part of life. Have you ever wondered how many times you do each of these in a year? How would you go about counting something like that? Trivial you say? Think again, for once more, we are close to touching upon timeless principles: touching upon those "special cases" of David Hilbert's, if you will—the ones containing all the germs of generality.

In this chapter, we will find that figuring out how to conveniently count repetitive actions or situations leads us onto the trail of an entirely new way of reckoning—called multiplication. We begin by considering the following:

A. A bookstore receives a regular shipment of 15 boxes of books every week. How many boxes are delivered in one year?
B. The setup for a graduation ceremony has 30 rows each with 45 chairs. How many guests can be seated at the ceremony?
C. What is the total amount paid back on a signature loan of 36 months where each payment is $115?

D. A 2001 Chevrolet Impala gets 25 miles per gallon. If it has a gas tank which holds 17 gallons, how many miles can it travel when full?

E. A country has a population of 26,784,000 (approximately the same number of people as Nepal in 2011). If the average per capita income is $17,200, what is the combined income in one year for all of its citizens?

These five problems have something in common with our questions about weekly routines; and most people would answer that each can be solved by multiplication. But what exactly does that mean? If a child asked you what multiplication is about, could you explain it? If we knew nothing of multiplication and could only add and subtract, would we still be able to find the answers to these questions?

While each of these problems deals with a different issue, they can be solved by addition alone—provided we add together the number involved in the repetitive act the required number of times. For question:

A. The repetitive act is a shipment of 15 boxes every week for a year (52 weeks in a year); for the answer we must add fifty-two 15s together.

B. The repetitive act is 45 chairs in each of the 30 rows of chairs in the auditorium; for the answer we must add thirty 45s together.

C. The repetitive act is a payment of $115 for each of the 36 months of a loan; for the answer we must add thirty-six 115s together.

D. The repetitive act is 25 miles for each of the 17 gallons of gas in the fuel tank; for the answer we must add seventeen 25s together.

E. The repetitive act is $17,200 for each of the 26,784,000 citizens in the country; for the answer we must add twenty-six million, seven hundred eighty four thousand 17200s together.

The need to repeatedly add or count in multiples occurs in numerous guises.

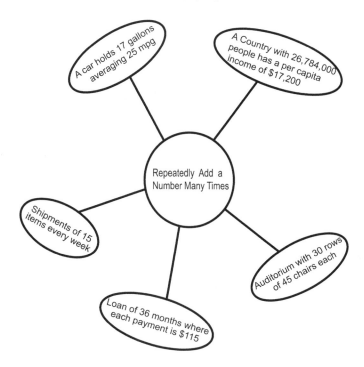

There are millions upon millions of other situations (an infinite nation of them if you will) that can be linked by the need to repeatedly add numbers together and this fact alone makes this process highly relevant. As with Amar, when he began studying circles: Whatever we discover in our study should have far-reaching, even universal, implications.

But first we need to give chase to the common thread running through all of these problems. What is it? Each of the problems involves a "certain number of objects" (boxes, chairs, miles, or dollars) repeatedly used a "certain number of times." These two numbers are what we must capture in representing this common thread. In other words, we must identify: the "number of times the repetitive act occurs" and the "number of objects involved in each repetitive act."

For the situation involving bookstore shipments the repetitive act occurs 52 times and 15 objects are involved in each repetitive act. Commonly used abbreviations linking these two numbers include: 52×15 or $52 \cdot 15$ or $(52)(15)$. Each of these reads as: add fifty-two 15s together or 52 times 15. In the same spirit, we could rewrite the "total combined income" question as 26,784,000 times 17,200 or $26,784,000 \times 17,200$. These shorthand abbreviations are written representatives of a new process—one that is different and distinct from simple addition or simple subtraction. We will christen the method with the name "multiplication."

While we may jump for joy over this christening, with its shorthand notation, it really isn't very helpful in and of itself. That is, if the only way to calculate 26,784,000 × 17,200 is to literally take twenty-six million, seven hundred and eighty-four thousand 17,200s and add them together, then using this abbreviation hasn't really given us anything new from the standpoint of saving us work—we still have to do all of the additions to solve problems.

Solving the problem this way is simply out of the question: There are 26,784,000 seconds in 310 days, and it most surely would take a longer time than this to simply count from 1 to 26,784,000—even if that is all one did 24/7.[2] Since it takes more time to add than to count, it stands to reason that performing these many repeated additions of 17,200 in a direct manner (without any shortcuts) would take a great deal longer than 310 days—even using our modern method of doing addition.

Nevertheless this type of problem remains and needs to be solved. What are we to do? We know many of our ancestors got around this problem by using devices such as the abacus to help them multiply. But we have also shown that the HA numerals are a direct capture of a certain type of abacus in writing. Is it possible to now take advantage of this and avoid actually scribbling down a string of numerals the required number of times—making convenient multiplication in writing possible? If so, this would be a bonanza indeed.

In this chapter, we begin an investigation into whether this is possible or not, and will follow G. K. Chesterson's lead by looking at the familiar territory of multiplication as if we were strangers in a new land.

Egyptian Multiplication

Our task looks daunting. We have to engineer a way to compute on paper (or some other writing medium) the net effect of doing lots of additions, potentially numbering in the millions, without actually performing all of those additions. Is such a path even possible? Since the abacus already does such an adequate job, why should we even care? Many civilizations didn't—the methods using a device worked for them and that was good enough.

Nevertheless, performing multiplication in writing is not something we should take lightly. Language writing opens up to us whole new ways of looking at the world and so too might the ability to multiply in writing.

How should we proceed?

We start by asking a simple question of multiplication—does the order in which we multiply make a difference in our answer? We know that order doesn't matter in addition; for example, we get 13 whether we add 4 + 9 or 9 + 4. This quality of addition is called the commutative property. Subtraction evidently does not have this property since 6 − 2 = 4 and 2 − 6 does not. The latter subtraction doesn't even make sense for whole numbers, that is, when it is applied to collections such as a group of people or a collection of cell phones.[3]

It would be fabulous if the order in which we multiply two numbers gave the same answer, since it would mean that 825 × 2 gives the same value as 2 × 825, which would imply that adding "825 twos" together is equivalent to adding "2 eight hundred and twenty fives" together. The latter requires the addition of only two numbers! Put another way, if the order in which we multiply doesn't change our answer, then we can find the net effect of adding 825 numbers together by simply changing our viewpoint and adding together only 2 values.

We address the question of order in multiplication by considering an example. We use coin numerals for conceptual illumination: Does 4 × 2 = 2 × 4?

These are equivalent. In fact, if we make the columns in B into rows and the rows into columns, we obtain A.

The process of converting rows to columns and vice versa is called "taking the transpose." Thus the transpose of the result in B is equal to the result in A. Similarly, the transpose of A is equal to B. We clearly see that 4 × 2 = 2 × 4 = 8.

This example illustrates what is true in general—the order in which we multiply two numbers makes no difference in the final answer. The order in which we repeatedly add or multiply, however, does make a big difference in how convenient a time we have in obtaining that answer. Using this basic property in the per capita income situation, we can find the value of 26,784,000 × 17,200 (requiring that we add 26,784,000 numbers together) by simply reversing the order and calculating 17,200 × 26,784,000 instead (requiring that we only add 17,200 numbers together).

The equality of these multiplications quite naturally connects the total income of two radically different nations, one with a large population of more than 26 million where the average citizen has an annual salary of $17,200 (below what is considered the poverty line for an American family of four), with a very rich nation of only 17,200 citizens where each earns on average $26,784,000 per year.

Taking a situation that requires more than 26 million additions and transforming it to one that requires 17,200 is no small feat. Unfortunately, this is still way too many to make brute force repeated addition a practical way to solve problems that involve many repetitive acts.

Remarkably in the time of the pharaohs, the ancient Egyptians developed a systematic way to perform fewer additions in writing. Their technique sheds light on an important property of multiplication so we will discuss it in some detail here.

The method is based on the principle of doubling and requires forming two rows. The values in the top row are obtained by starting with the number one in the first cell, and repeatedly doubling the result as we move across the row—thus, doubling 1 yields 2, doubling 2 yields 4, doubling 4 yields 8, and so on. We list the results in the top row up to 1024:

Doubling Table

1	2	4	8	16	32	64	128	256	512	1024

The bottom row will change values depending on the number repeatedly added. Let's look at this for the case where the number is 5. The numbers in the bottom row are obtained by starting with the number 5 in the first cell and doubling as we move across the row.

Doubling Table for 5

1	2	4	8	16	32	64	128	256	512	1024
5	10	20	40	80	160	320	640	1280	2560	5120

This is a times table of sorts for 5, since for example 5 × 8 = 40 and the 40 lies below the 8. Believe it or not, from just the 11 entries in the bottom row we will be able to calculate the value of 5 times any number between 1 and 2047.[4] The two examples of 14 × 5 and 642 × 5 show how this works:

Doubling Table (14 × 5)

1	2	4	8	16	32	64	128	256	512	1024
5	10	20	40	80	160	320	640	1280	2560	5120

To use the table to compute 14 × 5, we first scan the top row for numbers that add up to 14 and find that 8 + 4 + 2 = 14.

Next, we add the numbers in the lower row which lie below each of these to obtain: 40 + 20 + 10 = 70. We conclude that 14 × 5 = 70.

Expanding this sum illustrates conceptually what is happening.

40	+	20	+	10	=	70
(5+5+5+5+5+5+5+5)		(5+5+5+5)		(5+5)	=	70
8 fives	+	4 fives	+	2 fives	=	14 fives

To find 642 × 5, we scan the top row of the table for numbers that add up to 642 and find that 512 + 128 + 2 = 642.

Doubling Table (642 × 5)

1	2	4	8	16	32	64	128	256	512	1024
5	10	20	40	80	160	320	640	1280	2560	5120

We add the numbers in the lower row which lie below each of these to obtain: 2560 + 640 + 10 = 3210. We conclude that 642 × 5 = 3210. Note that 2560 contains 512 fives and 640 contains 128 fives and 10 contains two fives. Hence their sum contains 642 fives which is our goal.

To see the generality of this method we now construct a doubling table for 45 and use it to compute 1584 × 45:

Doubling Table for 45

1	2	4	8	16	32	64	128	256	512	1024
45	90	180	360	720	1440	2880	5760	11520	23040	46080

We scan the top row to find the numbers adding up to 1584 and find that 1024 + 512 + 32 + 16 = 1584.

Doubling Table (1584 × 45)

1	2	4	8	16	32	64	128	256	512	1024
45	90	180	360	720	1440	2880	5760	11520	23040	46080

We add the corresponding numbers in the bottom row to obtain: 46080 + 23040 + 1440 + 720 = 71280. We conclude that 1584 × 45 = 71280.

From the definition of multiplication as repeated addition, we can find 1584 × 45 by either adding 1584 (45s) together or by reversing the order and shortening it to adding 45 (1584s). Adding together forty-five numbers is still more work than we would like to do. Using the procedure from ancient Egypt allows us to find the answer (after the table has been constructed) by adding together only 4 numbers. That's right, only 4! Pretty clever of the ancients! The amount of effort saved is even more substantial when larger numbers are involved.

In illustrating the Egyptian method here, we have used our modern day HA numerals. The Egyptians of course did not know these numerals and would have used some other set. It turns out, in fact, that the Egyptians had several types of numerals, including the hieroglyphic numeral system discussed earlier as well as a cursive form of numerals called hieratic. They evidently used hieratic numerals when performing the doubling method. The doubling method is demonstrated in one of the oldest known surviving documents on mathematics—the Rhind Papyrus. Based on an even earlier document, the work is estimated to have been written around 1650 BCE by a scribe named Ahmes.[5]

The Egyptian method convincingly demonstrates that we can obtain answers to multiplication problems in systematic ways that don't require doing all of the additions in writing. In fact, "17,200 × 26,784,000" can now be solved by many within 15 to 25 minutes of using it. The simple and ancient method of doubling, combined with the commutative property of multiplication, has accelerated a process that in its original form would have taken several hundred days to perform using brute force repeated addition to one that can now be completed on a few sheets of paper and in less time than it takes to watch a single rerun of *The Cosby Show*. That's an exceptionally cool thing.

The Distributive Property

Crucial to the success of the Egyptian method is the distributive property of multiplication over addition. To illustrate how the doubling method works, we will first break down a couple of multiplications by 11. We start with 11 × 5; 11 × 5 means to add 5 together 11 times:

$$5 + 5 + 5 + 5 + 5 + 5 + 5 + 5 + 5 + 5 + 5$$
$$11 \ fives$$

We have a lot of freedom in how we choose to add these eleven fives. We can do it by first adding five of the 5s together in one group and six of the 5s in the second group and then combining the results together to obtain 55:

$$5 + 5 + 5 + 5 + 5 \qquad + \qquad 5 + 5 + 5 + 5 + 5 + 5 \qquad = \quad 11 \times 5$$
$$\text{5 \textit{fives}} \qquad\qquad\qquad \text{6 \textit{fives}}$$

$$5 \times 5 \qquad + \qquad 6 \times 5 \qquad = \quad 55$$
$$25 \qquad\qquad\qquad 30$$

We can also arrange the numbers in the following two ways as well:

$$5 + 5 + 5 + 5 \qquad + \qquad 5 + 5 + 5 + 5 + 5 + 5 + 5 \qquad = \quad 11 \times 5$$
$$\text{4 \textit{fives}} \qquad\qquad\qquad \text{7 \textit{fives}}$$

$$4 \times 5 \qquad + \qquad 7 \times 5 \qquad = \quad 55$$
$$20 \qquad\qquad\qquad 35$$

or

$$5 + 5 \qquad + \qquad 5 + 5 + 5 + 5 + 5 + 5 + 5 + 5 + 5 \qquad = \quad 11 \times 5$$
$$\text{2 \textit{fives}} \qquad\qquad \text{9 \textit{fives}}$$

$$2 \times 5 \qquad + \qquad 9 \times 5 \qquad = \quad 55$$
$$10 \qquad\qquad\qquad 45$$

Given that 11 can be broken up into 5 + 6, 4 + 7, or 2 + 9, the above suggests that we can calculate "11 × 5 = 55" directly or by using any of these subdivisions of 11. We will get 55 no matter which route we choose to take.

This is where the doubling table for 5 in the Egyptian method comes in (see the following abbreviated table):

Abbreviated Doubling Table for 5

1	2	4	8
5	10	20	40

The fact that 11 equals 8 + 2 + 1 means that we can break up adding the eleven 5s into groups that are already listed in the table. This allows us to swiftly conclude that:

$11 \times 5 = (5 + 5 + 5 + 5 + 5 + 5 + 5 + 5) + (5 + 5) + \ (5) \ = 40 + 10 + 5 = 55$
11 *fives*   8 *fives*   2 *fives*   1 *five*

Similarly the fact that $8 + 4 + 2 + 1$ equals 15 means that we can break up adding the fifteen 5s into groups that are also already given in the table. We swiftly conclude again that:

$15 \times 5 = (5 + 5 + 5 + 5 + 5 + 5 + 5 + 5) + (5 + 5 + 5 + 5) + (5 + 5) + \ (5)$
15 *fives*   8 *fives*   4 *fives*   2 *fives*   1 *five*
$=$ 40 $+$ 20 $+$ 10 $+$ 5 = 75

Critical to making this doubling procedure work is that the numbers 1, 2, 4, and 8 supply complete coverage of every whole number between 1 and 15 (i.e., any number between 1 and 15 can be written as a sum involving only 1, 2, 4, and 8):

Coverage of Whole Numbers from 1 through 15

1 = 1	**5** = 4 + 1	**9** = 8 + 1	**13** = 8 + 4 + 1
2 = 2	**6** = 4 + 2	**10** = 8 + 2	**14** = 8 + 4 + 2
3 = 2 + 1	**7** = 4 + 2 + 1	**11** = 8 + 2 + 1	**15** = 8 + 4 + 2 + 1
4 = 4	**8** = 8	**12** = 8 + 4	

This implies that multiplying 5 by any number between 1 and 15 can be obtained by adding some combination of the four elements in the doubling table for 5. This can easily be extended to coverage of the whole numbers between 16 and 31 by simply appending another entry to the doubling table:

Appended Abbreviated Doubling Table for 5

1	2	4	8	16
5	10	20	40	80

The reader should note that all whole numbers between 1 and 31 can now be obtained from some combination from the top row of this expanded table (e.g., $27 = 16 + 8 + 2 + 1$ and $23 = 16 + 4 + 2 + 1$). This implies then that multiplying 5 by any number between 1 and 31 can be also obtained from some combination from the bottom row of the table (e.g., $27 \times 5 = 80 + 40 + 10 + 5 = 135$ and $23 \times 5 = 80 + 20 + 10 + 5 = 115$).

This process continues with the next two entries in the top row being 32 and 64 supplying expanding coverage of the whole numbers from 1 to 63 and then from 1 to 127, respectively. Continuing on ad nauseum, we eventually cover all of the whole numbers, meaning that we can in principle use the doubling table to calculate any multiplication involving the number five (see www.howmathworks.com for more sample problems and explanation of many of the examples given in this chapter).

There is nothing special about five. These methods will work for any whole number. The sums in the top row in these tables remain. The only change would be to replace the doubling of 5 in the bottom row by the doubling of the number in question. Hence, the Egyptian method gives us a complete recipe for how to multiply any two whole numbers in writing. This recipe enables us to get around the issue of having to perform all of the repeated additions. It is a stunning achievement of ancient Egyptian mathematics!

In spite of its success, however, this method is not the main event here. The high point of our discussion on multiplication will center on how it works using the "alphabetic" features of the HA system. For it is here that the symmetries of the ancient Indian system get fully unleashed— elevating the ability of human beings to circumvent repeated addition in writing to a high art.

Multiplication in Place-Value Systems

Place-value multiplication is devastatingly effective on multiple fronts. If the Egyptian method were likened to conventional explosives, then multiplication using the alphabetic features of HA numerals is nuclear. The HA conquest of multiplication depends on these key components:

1. The distributive property of multiplication over addition
2. The existence of convenient footholds into which we can naturally decompose any number
3. The symmetry of the positional system

To take full advantage of these three components requires that we:

A. Make multiplication alphabetic (as we did with addition) by constructing a "times" table.
B. Devise an efficient algorithm (that takes advantage of these alphabetic features) which is compact and straightforward to learn.

We have discussed the distributive property and also the existence of convenient footholds in the case of the Egyptian scheme (the entries in the top rows of the doubling tables). In the HA system, the convenient footholds are the place values, themselves, corresponding to 1, 10, 100, 1000, 10,000, and so on. We are going to use the footholds here a bit differently than in the Egyptian case.

Alone, the footholds corresponding to the place values cannot be combined in the Egyptian way. What the place values offer, however, is a much more natural decomposition of a number. This decomposition is, of course, their raison d'être. Before beginning, it will prove useful to introduce two hybrid or mixed numeral representations that will help provide critical conceptual understanding throughout our discussion of both multiplication and division. The decomposition of 625 illustrates these representations:

Coins: 625 =   6 (H)   +   2 (T)   +   5 (I)
Hybrid

Abaci Rods: 625 = 6 x | | | +   2 x | | +   5 x |
Hybrid

HA Script: 625 = 6 x 100   +   2 x 10   +   5 x 1

Multiplying a place value by a given number is very straightforward as the multiplications of 1, 10, and 100 by 3 show:

3 x 1 = 3 x | = | + | + | = | = 3

3 x 10 = 3 x | | = | | + | | + | | = | | = 30

3 x 100 = 3 x | | | = | | | + | | | + | | | = | | | = 300

For each calculation notice that the same thing happens—multiplication by 3 converts one bead (●) into three beads (● ● ●), the only difference being the location of the rod on which it occurs.

For 3 × 1, this conversion occurs on the rod in the ones place. For 3 × 10, this conversion occurs on the rod in the tens place. For 3 × 100, this conversion occurs on the rod in the hundreds place. This identical behavior in different locations again illustrates the symmetry of our positional system. It is also worth pointing out that multiplication by 3 has no effect on the vacant rods (represented by a zero in HA script), only on the rod which contains a bead—yielding three beads plus the vacant rods. In HA notation, this translates to a value of 3 followed by the number of zeros in the place value. Using this same observation we readily conclude that 3 × 10000 will yield a 3 followed by four zeros or 30000.

There is nothing special about 3; the same result holds for any number. Namely, multiplying any number by the place values (1, 10, 100, 1000, etc.), yields a value starting with the number followed by the quantity of zeros in the particular place value. Thus 12 × 10000 gives 12 followed by four zeros or 120000.

We have discovered a fundamental pattern from observing the behavior of the abacus rods for the multiplications: 3 × 1, 3 × 10, and 3 × 100. There are dozens of other fundamental patterns to be revealed. Let's look for the one yielded in the multiplications of: 4 × 3, 4 × 30, and 4 × 300 (we now add vertically for increased clarity):

The pattern this time is:

For 4 × 3, the situation is played out on rods one and two. For 4 × 30, the situation is played out on rods two and three. For 4 × 300, the situation is played out on rods three and four. In HA script, this fundamental pattern reads as 4 × 3 =12. Once this is known, we can handle any situation involving 4 times a number that has 3 as its only nonzero digit. We do this by simply leading with 12 followed by the correct amount of zeros (e.g., 4 × 3000000 = 12000000).

The fundamental pattern for 6 × 7 is:

For 6×7, this pattern is played out on rods one and two, while for 6×70, this pattern is played out on rods two and three and so on. In HA script, this fundamental pattern reads as $6 \times 7 = 42$. From this we can handle any situation involving 6 times a number that has 7 as its only nonzero digit. We do this by simply leading with 42 and following it by the correct amount of zeros (e.g., $6 \times 7000 = 42000$).

There are more jewels to be uncovered here (than might first appear on the surface) from the knowledge of the two fundamental patterns for $4 \times 3 = 12$ and $6 \times 7 = 42$. We can now project these patterns to handle many, many more complicated multiplications. Let's see this play out on the multiplication of 33×4—using the basic pattern that $4 \times 3 = 12$, the distributive property, and the knowledge that reversing the order in which we multiply doesn't change our answer:

$$33 \times 4 \quad = \quad 30 \times 4 \quad + \quad 3 \times 4 \quad = \quad 4 \times 30 \ + \ 4 \times 3$$
Add thirty three   Add thirty   Add three
4s   4s   4s

Using both abacus rod and HA script in parallel yields:

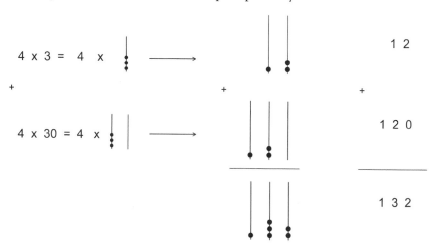

This gives $33 \times 4 = 132$. We easily extend this to the multiplication of 333×4 by writing this as:

$333 \times 4 \ = \ 300 \times 4 \ + \ 30 \times 4 \ + \ 3 \times 4 \ = \ 4 \times 300 \ + \ 4 \times 30 \ + \ 4 \times 3$

Add 333   *Add 300*   *Add 30*   *Add 3*

4s   4s   4s   4s

and doing the same thing as above we obtain $333 \times 4 = 12 + 120 + 1200 = 1332$. Try it on the rods yourself.

In the same vein, the fundamental pattern $6 \times 7 = 42$ allows us to calculate 777×6 as follows:

$777 \times 6 \ = \ 700 \times 6 \ + \ 70 \times 6 \ + \ 7 \times 6 \ = \ 6 \times 700 \ + \ 6 \times 70 \ + \ 6 \times 7$

Add 777   *Add 700*   *Add 70*   *Add 7*

6s   6s   6s   6s

which gives $777 \times 6 = 42 + 420 + 4200 = 4662$. This is how it plays out on abacus rods:

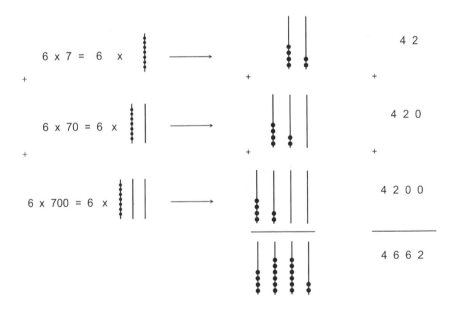

Simply knowing the fundamental pattern that $6 \times 7 = 42$ gives us the ability to multiply 6 by any number involving only 7s and 0s (e.g., 7777×6, 70707×6, 77700077×6). For example, the multiplication 70707×6 looks like:

$$70707 \times 6 = 70,000 \times 6 + 0000 \times 6 + 700 \times 6 + 00 \times 6 + 7 \times 6 =$$

$$6 \times 70,000 + 6 \times 0 + 6 \times 700 + 6 \times 0 + 6 \times 7 =$$

$$420000 + 0 + 4200 + 0 + 42 = 424242$$

Knowing that 4×3 is 12 will yield a similar result for multiplication by 4 of any number containing only 3s and 0s.

Thus the extent to which we can project our knowledge when we know a fundamental pattern is truly amazing. We may use this knowledge to literally multiply infinitely many more numbers. That is a lot of bang from a single pattern. In principle, if we know all of the fundamental patterns, we can project our knowledge even more (how far?). Is it possible to know all of the fundamental patterns?

The fundamental patterns are those that occur when we multiply the ten digits among themselves (e.g., 9×5, 4×7, 7×6). Since the number of such products is 100, we are yet again (for a third time) presented with the opportunity to make an operation alphabetic. If we learn and memorize these 100 fundamental patterns, it will be possible in principle to compute any product of the form (a single digit) × (any sized number).

The following tables show the 100 fundamental patterns in abacus rods (the tables reads as A × B):

Abacus Rod Multiplication Table

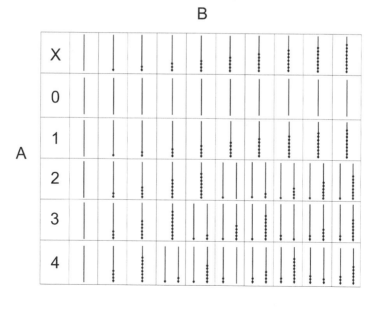

Abacus Rod Multiplication Table (Continued)

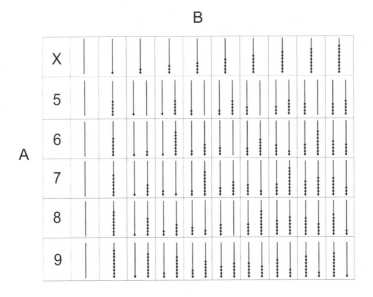

The power of this table can be shown in the calculation of 235×7:

$$235 \times 7 \ = \ 200 \times 7 \ + \ 30 \times 7 \ + \ 5 \times 7 \ = \ 7 \times 200 \ + \ 7 \times 30 \ + \ 7 \times 5$$

Add 235	*Add 200*	*Add 30*	*Add 5*
7s	7s	7s	7s

The fundamental patterns from the table are:

$7 \times$ | $=$ | ; $7 \times$ | $=$ | ; $7 \times$ | $=$ |

$(7 \times 2 = 1\ 4)$; $(7 \times 3 = 2\ 1)$; $(7 \times 5 = 3\ 5)$

Using these patterns and translating the situation to the proper location yields:

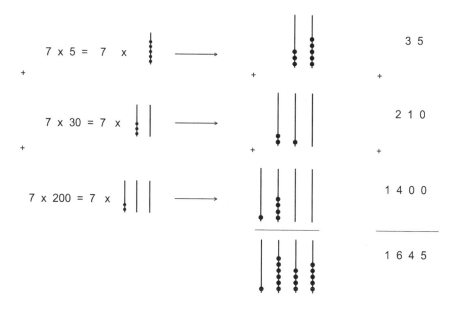

Thus $235 \times 7 = 1645$.

This method can be used to multiply a single-digit number by any sized number. To multiply a single digit by numbers with more than three digits (for instance, 5649×7), we simply add more rods to accommodate the larger place values and use the fundamental patterns given in the abacus rods table. This recipe using the abacus rods gives us a systematic method for handling all multiplications involving a single digit. Unfortunately, it is still too bulky for our tastes and needs to be further simplified.

As before with addition and subtraction, we can directly translate the abacus rod multiplication table into a table involving only HA script. The resulting table is probably the most well-known mathematical table of all and is called the multiplication table or times table:

Multiplication Table (or Times Table)

x	0	1	2	3	4	5	6	7	8	9
0	0	0	0	0	0	0	0	0	0	0
1	0	1	2	3	4	5	6	7	8	9
2	0	2	4	6	8	10	12	14	16	18
3	0	3	6	9	12	15	18	21	24	27
4	0	4	8	12	16	20	24	28	32	36
5	0	5	10	15	20	25	30	35	40	45
6	0	6	12	18	24	30	36	42	48	54
7	0	7	14	21	28	35	42	49	56	63
8	0	8	16	24	32	40	48	56	64	72
9	0	9	18	27	36	45	54	63	72	81

Armed with this table, we are able to multiply any number by a single digit in the manner of the previous examples, except now it can be done purely in script. The multiplication of 8432×8 shows how this is done:

$$8432 \times 8 = 8000 \times 8 + 400 \times 8 + 30 \times 8 + 2 \times 8 =$$

$$8 \times 8000 + 8 \times 400 + 8 \times 30 + 8 \times 2$$

The fundamental patterns from the HA-table are:

$8 \times 8 = 64; \quad 8 \times 4 = 32; \quad 8 \times 3 = 24; \quad 8 \times 2 = 16$

Translating the fundamental patterns to the proper location yields:

(tens and ones places)	8×2	=	16
(hundreds and tens places)	8×30	=	240
(thousands and hundreds places)	8×400		3200
(ten thousands and thousands places)	8×8000		64000
	Adding these yields:		67456

Thus $8432 \times 8 = 67456$.

Conclusion

Who would have guessed that tracking the total number of boxes shipped to a bookstore in a particular year or figuring out the total amount of money owed back on a signature loan of 36 months, would have anything to do with expressing timeless principles in nature? But it turns out that they do. Multiplication and its generalizations appear throughout the whole of mathematics and science—making their presence felt, even in many of the most esoteric attempts by scientists to describe symbolically how the world works (e.g., E = mc^2 or Energy = [the mass] × [the speed of light] × [the speed of light]).[7] In mathematics, by placing our fingers on a given problem, no matter how trite or pedestrian it apparently seems, we may end up measuring the pulse of the universe.

And it all begins here, with the type of ordinary problems we have discussed in this chapter. It is not unlike the physical scenario involving a tiny stream flowing out of the misty confines of a lake in northern Minnesota that eventually grows into a veritable torrent that becomes the world class Mississippi-Missouri River system ranking among the largest in the world.

Hyperbole aside, we are now midstream in our campaign to successfully grapple with multiplication. We have made significant strides in this chapter—acquiring the capacity, thanks to the Egyptians, to solve all of the originally stated problems (and infinitely more) by systematically sidestepping in some cases thousands of additions. Knowing we can do this gives the new operation of multiplication teeth as it means that we can work completely within the abbreviated form to more conveniently handle such problems. But our task is far from complete—much still needs to be simplified. This simplification is the focus of the next chapter.

7

Dance of the Digits

Dancing in all its forms cannot be excluded from the curriculum of all noble education: dancing with the feet, with ideas, with words, and, need I add that one must also be able to dance with the pen?

—Friedrich Nietzsche, German philosopher,
poet, classical philologist[1]

IF THIS BOOK REPRESENTS a journey passing through mountainous terrain, then this chapter represents one of the highest pinnacles that we shall attain. With many peaks, the toughest part of the climb is often the last part. In this vein, this chapter involves a bit more symbolic manipulation than some of the others—but it is still within your purview. I urge the less mathematically inclined of you to stay the course here. We have covered a lot of conceptual terrain to get to this point and now have placed within sight a much deeper understanding of how the modern methods of multiplication taught in the schools really work. Let us not now shirk from the symbols when we are so close. Hopefully you will find the scramble up this last segment of the trail illuminating (if needed, please visit www.howmathworks.com to see more examples).

The major focus of the previous chapter was to show that it is possible, in writing, to solve problems that require repeated addition by actually getting around doing all of those additions on paper. This allows us to significantly reduce the time it takes to find solutions in these situations—once more placing before us the universal idea of substituting one thing for another to gain decisive advantages. Due to its widespread occurrence, the name "multiplication" was given to this way of reckoning.

While much progress has been made in taming this operation (the Egyptian method of doubling, for instance), our times table methods in HA script are still not complete. Presently, we can only handle a restricted class of problems—namely, those involving an arbitrarily long number multiplied by a single digit. Computing such multiplications, in principle, has been completely solved and can usually be performed much faster, with times tables, than using the Egyptian technique. But there are many, many other multiplications where both numbers have two or more digits (the five original problems of the last chapter, for instance) and the time has come to learn how to deal with these.

Additionally, we still need to see if the entire process can be miniaturized further into a single diagram in a manner similar in spirit to the compact algorithms we have for addition and subtraction.

Our goals then are this: to develop a general recipe (in HA script) that works for the multiplication of any two whole numbers, not just the special cases of the previous chapter, and then to stylishly simplify it. We will accomplish both in this chapter, and in the demonstration will see the overwhelming power inherent in the symmetry of the HA notation. This symmetry will give us the ability to take the repeated addition of a number and radically transform it into something completely different, into something that in its most expressive form can be likened to a drill team of digits marching to well-scripted rules—a veritable *dance of the digits*, if you will. This transformation, this poetry of the diagrams, allows for the operation of multiplication, which is significantly harder than either addition or subtraction, to be wrestled out of the hands of the specialist and laid at the doorstep of elementary school-aged children.

Carving Up the Numerals

The plan of attack, when multiplying two numbers in HA script, is to "carve up" the multiplication or product in a way that allows us to unleash the times table on it. When the product is of the form "two or more digits × a single digit," as in the last chapter, we break up the larger number to accomplish this (e.g., in 62×7 we split up the 62 [as $60 + 2$] to get $60 \times 7 + 2 \times 7$). However, if both numbers in the product contain two or more digits, we have some freedom in how we choose to decompose. For example, when computing the product 62×37, we can choose to carve up the 62 first or the 37 instead. To successfully handle these cases requires that we relocate the trailing zeros in the product.

If a numeral ends in one or more zeros, we will call these zeros "trailing zeros." For example, 50 has one trailing zero and 500 has two trailing zeros. The numeral 100020 has four total zeros, but only the one at the end qualifies as a trailing zero. In HA script, trailing zeros are like free agents which can roam at will from the tail of one numeral in the product to the tail of the other. The following all show that while the total number of trailing zeros is conserved in the product (before completing the multiplication, that is), they can move freely between either numeral:

One Trailing Zero: $60 \times 1 = 6 \times 10 = 60$

Two Total Trailing Zeros: $60 \times 10 = 6 \times 100 = 600 \times 1 = 600$

Four Total Trailing Zeros: $600 \times 700 = 60 \times 7000 = 6000 \times 70 = 6 \times 70000 = 420000$

Five Total Trailing Zeros: $54000 \times 7200 = 54 \times 7200000 = 5400 \times 72000 = 5400000 \times 72 = 388800000$

In terms of our viewpoint of multiplication as repeated addition, these seemingly innocuous shifts of zero are actually tremendous simplifications. They have far greater impact on reducing the number of additions required than even does simply reversing the order of the multiplication. For instance, 600×700 means to take six hundred 700s and add them together. But by shifting zeros, we can swiftly rewrite this as 6×70000 and find the answer by knowing only that 6×7 is 42 and then putting the four zeros after it to obtain 420000—meaning in this case, that the collective effect of hundreds of additions has been rerouted to a single calculation that can be completed in less than ten seconds! These shifts play out on abacus rods as shown:

$$600 \times 700 = 600 \times \quad = 6 \times \quad = $$

$$600 \times 7\,0\,0 = 6 \times 7\,0\,0\,0\,0 = 4\,2\,0\,0\,0\,0$$

Exploiting the free agent properties of trailing zeros will prove to be pivotal to all that follows in that we can reposition them to convert multiplications between any two numbers, into multiplications found in the multiplication table. We just have to learn how to incorporate all of the shifts in such a way that everything aligns correctly—historically, no small task. If there are still any lingering doubts on the significance of the number zero (and the symbol representing it) to how we perform arithmetic, please let them cease here.

Nontrailing zeros, on the other hand, are bound and cannot be moved around at will since doing so gives different answers. For example, the zeros in the product, 7202×6008, are nontrailing and cannot be shuffled around: $7202 \times 6008 \neq 7220 \times 6800$.

To get the ball rolling let's look at how the multiplication of "two digits $\times$ two digits" plays out by walking through "62×37":

Carving Up Sixty-Two First (60 + 2)

62×37	$=$	60×37	$+$	2×37
Add 62		*Add 60*		*Add 2*
37s		37s		37s
		B	$+$	A

- A = 2×37:
  In this form we still can't directly engage the multiplication table. We must also split up the 37 (as 30 + 7) which gives: $2 \times 37 = 2 \times 30 + 2 \times 7$.
  Now we can directly engage the table.

The fundamental patterns from the times table are: $2 \times 3 = 6$; $2 \times 7 = 14$

Using these patterns gives: $2 \times 30 + 2 \times 7 = 60 + 14 = 74$.
And we have $2 \times 37 = 74$.
- B = 60×37:
  In the form 60×37, we can't directly engage the multiplication table. We must first reposition the trailing zero:

60×37	$=$	6×370
		Moving the
		trailing zero
		to the right

Decomposing the 370 (300 + 70) allows us again to engage the multiplication table:

$$6 \times 370 = 6 \times 300 + 6 \times 70 = 1800 + 420 = 2220$$

The fundamental patterns from the times table are: $6 \times 3 = 18$; $6 \times 7 = 42$.

Adding the result in *A* to the result in *B* vertically:

$$
\begin{array}{r}
74 \\
+ \ \underline{2220} \\
2294
\end{array}
$$

This gives $62 \times 37 = 2294$.

Carving Up Thirty-Seven First (30 + 7)

Since the order in which we multiply doesn't matter (i.e., $62 \times 37 = 37 \times 62$) we can read 62×37 from right to left (as 37 times 62). This interprets as adding together thirty-seven 62s:

62×37	=	60×30	+	62×7
Add 37		*Add 30*		*Add 7*
62s		62s		62s
		B	+	*A*

- A = 62×7:
  As before, we can't directly engage the multiplication table using the form 62×7. We need to also decompose 62 (60 + 2) which gives:

  $$62 \times 7 = 60 \times 7 + 2 \times 7 = 420 + 14 = 434$$

Thus $62 \times 7 = 434$.

The fundamental patterns used from the times table are: $6 \times 7 = 42$; $2 \times 7 = 14$

- B = 62×30:
  Similarly in the form 62×30, we can't directly engage the table. We must first reposition the trailing zero:

$$62 \times 30 \qquad = \qquad 620 \times 3$$

*Moving the
trailing zero
to the left*

Now we decompose the 620 (600 + 20) which gives:

$$620 \times 3 = 600 \times 3 + 20 \times 3 = 1800 + 60 = 1860$$

Thus $62 \times 30 = 1860$.

The fundamental patterns used from the times table are: $6 \times 3 = 18$; $2 \times 3 = 6$.

Adding the results in *A* to the result in *B* vertically:

$$
\begin{array}{r}
434 \\
+\ 1860 \\
\hline
2294
\end{array}
$$

This gives $62 \times 37 = 2294$.

We obtain the same value of 2294 whether we decompose the 62 first or the 37 first. Regardless of which number we choose to break apart first, we end up carving up the other one as well. Breaking up both numbers is necessary to fully engage the times table.

Trailing Zeros Go to Work

We are on the scent of something significant here. To get a better feel for exactly what that is let's give ourselves a longer piece of rope to play with this time by looking at the multiplication of 745×289:

Decomposing the 289 first:

745×289	$=$	745×200	$+$	745×80	$+$	745×9
Add 289		*Add 200*		*Add 80*		*Add 9*
745s		745s		745s		745s

Decomposing the 745 first:

745×289	$=$	700×289	$+$	40×289	$+$	5×289
Add 745		*Add 700*		*Add 40*		*Add 5*
289s		289s		289s		289s

To engage the multiplication table requires that we eventually take apart both numbers, the 745 next in the first case (to 745 = 700 + 40 + 5) or the 289 next in the second case (to 289 = 200 + 80 + 9). Once we have done both decompositions it then becomes a game involving only trailing zeros as indicated here:

$$
\begin{aligned}
745 \times 289 \quad &= 745 \times 200 = 700 \times 200 + 40 \times 200 + 5 \times 200 \text{ (1st row)} \\
&+ 745 \times 80 \ = 700 \times 80 \ + 40 \times 80 \ + 5 \times 80 \ \text{ (2nd row)} \\
&+ 745 \times 9 \ = 700 \times 9 \ \ + 40 \times 9 \ \ + 5 \times 9 \ \ \text{ (3rd row)}
\end{aligned}
$$

or,

$$
\begin{aligned}
745 \times 289 \quad &= 700 \times 289 = 700 \times 200 + 700 \times 80 + 700 \times 9 \\
&+ \ \ 40 \times 289 = \ \ 40 \times 200 + \ \ 40 \times 80 + \ \ 40 \times 9 \\
&+ \ \ \ \ 5 \times 289 = \ \ \ \ 5 \times 200 + \ \ \ \ 5 \times 80 + \ \ \ \ 5 \times 9
\end{aligned}
$$

Notice that each of the nine partial products in the first product (745 × 289) are present in the second product (*745 × 289*) as well. In fact, the rows in the first product become the columns in the second product (e.g., the three partial products in the first row of the first product, 700 × 200 + 40 × 200 + 5 × 200, become the first vertical column in the second product) and vice versa (this is the transpose again). As a result of this equivalence, we need only use one of the two diagrams as we continue (we choose the first one with the labeled rows).

Now that we have taken apart both numbers completely, it becomes possible to fully engage the multiplication table for every product by simply repositioning the trailing zeros:

$$745 \times 289 \quad = \quad \begin{array}{llll} (7 \times 2)\,0000 & + \;(4 \times 2)\,000 & + \;(5 \times 2)\,00 & \textit{(1st row)} \\ + \;(7 \times 8)\,000 & + \;(4 \times 8)\,00 & + \;(5 \times 8)\,0 & \textit{(2nd row)} \\ + \;(7 \times 9)\,00 & + \;(4 \times 9)\,0 & + \;(5 \times 9) & \textit{(3rd row)} \end{array}$$

Sum of Fundamental Products for 745×289

The fundamental products from the multiplication table are now: $\{7 \times 2, 4 \times 2, 5 \times 2, 7 \times 8, 4 \times 8, 5 \times 8, 7 \times 9, 4 \times 9, 5 \times 9\}$. Computing these gives:

$$745 \times 289 \quad = \quad \begin{array}{llllll} & 140000 & + & 8000 & + & 1000 \\ + & 56000 & + & 3200 & + & 400 \\ + & 6300 & + & 360 & + & 45 \end{array}$$

Adding these all up gives $745 \times 289 = 215{,}305$.

The systematic idea then is this: to fully break apart both numbers in a product, no matter how many digits, until we get down to single digits followed by trailing zeros. It is always possible to drill down a product this way in HA script. Once accomplished, it then becomes possible to fully engage the multiplication table (by simply repositioning trailing zeros) to find all of the partial products and then we complete the additions for the final answer.

There are patterns galore here and nowhere is that better demonstrated than in the "Sum of fundamental products for 745×289" expression. The following table makes these patterns explicit by listing the number of trailing zeros in each partial product entry with its accompanying coin denomination:

Zero Counts for 745 × 289

4 zeros (TTH)	3 zeros (TH)	2 zeros (H)	*1st Row*
3 zeros (TH)	2 zeros (H)	1 zero (T)	*2nd Row*
2 zeros (H)	1 zero (T)	No zero (I)	*3rd Row*

The pattern is the same whether we read the table horizontally or vertically (i.e., turning the rows into columns or vice versa yields the same arrangement). Is it possible to capitalize on these patterns? Most definitely!

Our next task then is to engineer a way to take advantage of these symmetries and translate this entire way of multiplying into a simpler, more convenient form.

Multiplication on Diagrams

The breakdown of 745 × 289 (into nine partial products involving the multiplication of single digits and trailing zeros) is general and will work in principle for the multiplication of any two whole numbers. And though it represents another triumph over brute force repeated addition (i.e., we can find 745 × 289 this way by doing far fewer than 289 additions), the method is still bulkier than we'd like and nowhere near as simple as it can be. We have seen that a very compact algorithm is possible in addition (shown here), and it is natural to ask if such a thing is possible with this type of multiplication.

Expanded addition versus compact addition (742 + 659):

Expanded Addition:

$$
\begin{array}{cccccccc}
 & (700 & + & 40 & + & 2) & & \\
+ & (600 & + & 50 & + & 9) & & \\
\hline
 & 1300 & + & 90 & + & 11 & = & 1401 \\
\end{array}
$$

Compact Addition:

$$
\begin{array}{r}
11 \\
742 \\
+ \ 659 \\
\hline
1401 \\
\end{array}
$$

To obtain a recipe which is similar in spirit to compact addition, we will work from the table listing the count of zeros (for 745 × 289). A major obstacle in trying to acquire conceptual leverage over HA procedures is that we use the same symbols repeatedly in different locations or place values. This makes it very easy to get confused and lose track of what is really going on (is that really a 4000 or should it be 400 and how does it align).

With HA numerals, the positions of digits within a given string of numerals are how we distinguish place values. Coin numerals, on the other hand, allow us to distinguish different locations or place values by using different denominations; meaning that the coins can act as tracking devices, if needed, and this can turn out to be conceptually very useful—especially for multiplication and division. Let's now exploit this to the hilt.

To keep everything tidy for our present purposes, we will represent the coin numerals without circles (including only the content of their value). Which means that throughout the rest of the chapter, we will characterize the coin numerals, (T) and (H), respectively, by the letters "T" and "H." Looking at the "Zero Counts Table for 745 × 289" and retaining only the coin information we obtain:

Zero Counts for 745 × 289—Coins Only

TTH	*TH*	*H*	*1st Row*
TH	*H*	*T*	*2nd Row*
H	*T*	*I*	*3rd Row*

Before analyzing this table further, it can be helpful to use this diagram as a road map of sorts for multiplication. A road map gives us a convenient and useful model of a three-dimensional landscape. Convenience comes from the fact that it is astronomically easier to lay a road map, say of the state of Texas, in our lap than it is to do so with the actual material state (all 268,000 square miles of it).

Utility comes from knowing that, even though a lot of information gets lost in representing such a landscape on small sheets of flat paper (i.e., we do not know what a region really looks like, smells like, or feels like by reading such a road map), key structural information, nonetheless, is still captured (items such as names and numbers of roads, which roads go through which towns, which roads connect to other roads, what are the main roads, and so on). In most cases, this information turns out to be precisely what drivers need to successfully navigate through a part of the country that they have never been in before. Exploiting road maps in this way is no trivial matter to any who have used them to travel.

Now in a similar fashion, these coin diagrams will provide useful insight into the world of multiplication—insights allowing us to "navigate" our way to powerful procedures for performing the operation.

Notice in the previous table that the denominations can be aligned if we look at them diagonally from right to left:

Demonination Alignment for Zero Counts—Coins Only (745 × 289)

TTH	TH	H
TH	H	T
H	T	I

This alignment is what we will build from to construct the first of the miniaturized methods allowing us to tame the multiplication of whole numbers. To make it work, requires we use the following table showing how the coin denominations (place values) multiply among themselves:

Multiplication Table for Coin Denomnation or Place-Values

x	I	T	H	TH
I	I	T	H	TH
T	T	H	*TH*[a]	TTH
H	H	TH	TTH	*HTH*
TH	TH	TTH	HTH	M

[a] $T \times H = 10 \times 100 = 1000 = TH$

We will now demonstrate how all of the pieces fit together by looking at the multiplication of 745 × 289 on the following sequence of grids:

A: Blank Grid

7 H   4 T   5 I

2 H
8 T
9 I

B: Place Values Written as Products

7 H   4 T   5 I

H x H	T x H	I x H
H x T	T x T	I x T
H x I	T x I	I x I

2 H
8 T
9 I

C: Place Values Multiplied (i.e., H × T = TH, etc.)

7 H   4 T   5 I

TTH	TH	H
TH	H	T
H	T	I

2 H
8 T
9 I

The sequence of steps A, B, C have gotten us back to the "Zero counts table for 745 × 289 – coins only" but this time in three steps, as opposed to the slower method in the previous section involving the repositioning of trailing zeros. Now we can multiply each of the individual digits directly on the table as follows:

D: Digits with Place Values Multiplied

7 H	4 T	5 I	
7 H x 2H = 14 TTH	4T x 2H = 8TH [a]	5I x 2H = 10 H	2 H
7 H x 8T = 56 TH	4T x 8T = 32 H	5I x 8T = 40 T	8 T
7 H x 9I = 63 H	4T x 9I = 36 T	5I x 9I = 45 I	9 I

[a] 4 T × 2 H = (4 × 2) (T × H) = 8 TH

E: Simplified Multiplications

7 H	4 T	5 I	
14 TTH	8 TH	10 H	2 H
56 TH	32 H	40 T	8 T
63 H	36 T	45 I	9 I

The fundamental products used to multiply 745 × 289 on the diagram here, {7 × 2, 4 × 2, 5 × 2, 7 × 8, 4 × 8, 5 × 8, 7 × 9, 4 × 9, 5 × 9}, match those used in the previous section. Note that the same denominations occur on the diagonals (reading from right to left). Adding like denominations or place values gives:

						14 TTH	140000
	8 TH	+	56 TH	=	64 TH	64000
10 H	+	32 H	+	63 H	=	105 H	10500
	40 T	+	36 T	=	76 T	760
					45 1	45

215,305

And the answer we obtain matches our result in the previous section for 745 × 289.

The "Sum of fundamental products for 745 × 289" diagram, which involved nine partial sums has now been *road-mapped* to nine entries in a table. The only difference between the two is appearance and how we get to them. The essential information is all the same with the same fundamental products from the times table being used in both cases. There is nothing that prevents all of the steps in the diagrams A–D from being compressed into the last diagram shown in E.

We demonstrate this by multiplying 452 × 78 on a single table (remembering to multiply both the digits as well as the place values):

4 H	5 T	2 I	
28 TH	35 H	14 T	7 T
32 H	40 T	16 I	8 I

The fundamental products used from the multiplication table are: $4 \times 7 = 28$, $5 \times 7 = 35$, $2 \times 7 = 14$, $4 \times 8 = 32$, $5 \times 8 = 40$, $2 \times 8 = 16$.

Adding like place values along the diagonals gives:

				28 TH	28000
35 H	+	32 H	=	67 H	6700
14 T	+	40 T	=	54 T	540
				16 1	16
					35256

Thus $452 \times 78 = 35256$.

Now we have the makings of a time and space saving procedure. We start it off by performing fundamental multiplications on a diagram. The entries in the diagram naturally separate the partial products by place values. After this, we simply pluck the entries off the table like fruit, adding them together, and we are done.

In spite of this simplification, the procedure is still not as elegant as it can be. Right now we have a hybrid method using both HA numerals and coin numerals. The coins have been used much like conceptual boosters to add insight. Are we now stuck with them or can we jettison them, à la NASA, and develop an algorithm that only uses HA script?

The Lattice Method

The time has come to let the numerals dance. It turns out that HA numerals can dance in many ways. The remainder of the chapter is devoted to the discussion of two such routines.

Let's begin by looking at the diagram corresponding to multiplying 23 × 56:

2 T	3 I	
10 H	15 T	5 T
12 T	18 I	6 I

Plucking the entries from each of the blocks and adding yields:

				10 H	1000
15 T	+	12 T	=	27 T	270
				18 I	18
					1288

Rather than adding the same place values to each other first, we can instead keep each entry from the table separate and add them vertically:

10 H	=	1000
15 T	=	150
12 T	=	120
18 I	=	18
		1288

Now notice that each of the hybrid entries on the left can be simplified first by carrying; that is, 10 "H" (ten hundreds) can be simplified to 1 "TH" (one thousand) and 12 "T" (twelve tens) can be simplified to 1 "H" and 2 "T" (one hundred, two tens). Let's see what happens if we put these changes in first before adding:

10 H	=	1 TH	0 H		
15 T	=		1 H	5 T	
12 T	=		1 H	2 T	
18 T	=			1 T	8 I
		1 TH	2 H	8 T	8 I

The answer matches what we had previously, as it should, since we are only reorganizing the sums and not changing the values. Now observe what happens on the diagram for 23 × 56 if we do these simplifications first:

	2	3	
	10 **H** = 1 TH / 0 H	15 **T** = 1 H / 5 T	5
	12 **T** = 1 H / 2 T	18 **1** = 1 T / 8 I	6

Judiciously reorganizing in each cell gives:

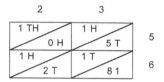

Things are aligned quite nicely now. If we draw a diagonal in each rectangle, we can see just how aligned the coin values have become.

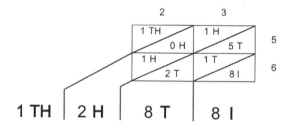

The same denominations lie in the same lanes. In this format, we will be able to simply add the numbers in each lane to obtain the total amount we have for each place-value denomination. Extending the lanes and adding yields:

If we read the grid from right to left, we observe that:

The ones denomination is in the first lane.
The tens denomination is in the second lane.
The hundreds denomination is in the third lane.
The thousands denomination is in the fourth lane.

Given that the lanes become place values when extended, the above translates to:

The ones sum is in the first location or the ones place.
The tens sum is in the second location or the tens place.

The hundreds sum is in the third location or the hundreds place.
The thousands sum is in the fourth location or the thousands place.

This perfect alignment let's us completely drop the coin tags and simply work with the numerals themselves:

23 × 56 (No Coin Tags)

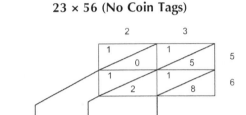

The advantage of this last procedure is that it uses HA symbols alone—the conceptual coin tags have been jettisoned! Our goal of obtaining a miniaturized and efficient algorithm using purely HA script has been achieved.

Let's demonstrate this procedure by multiplying 852 × 74:

852 × 74 (Blank Lattice)

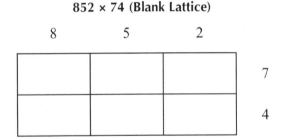

Placing the diagonal lines in for each cell gives:

852 × 74 (Blank Lattice & Diagonals)

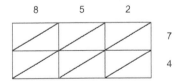

The products we have to consider from the times table of chapter 6 are:

Top row: 8 × 7 = 56, 5 × 7 = 35, 2 × 7 = 14
Bottom row: 8 × 4 = 32, 5 × 4 = 20, 2 × 4 = 8

Placing these products into their proper positions in the lattice yields:

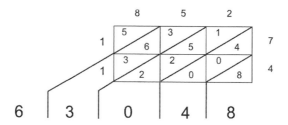

Extending the lanes, adding (from right to left), and carrying (indicated by the 1s outside the squares) in formation gives:

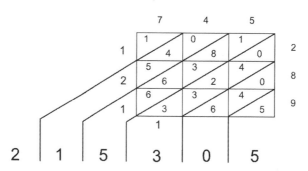

Thus $852 \times 74 = 63048$.

This can all be done on a single diagram. Doing so for our signature multiplication of the chapter (745×289) gives:

And we have $745 \times 289 = 215305$. Compare this to all of our earlier methods of calculating this product.

This then is the lattice method of multiplication. It allows for the complete transformation of the problems of repeated addition into a procedure in writing that involves placing numbers into formation, via a multiplication table, and simply adding them along the lanes. It is as if the numbers have become part of a dance troupe routine, a routine in which we are able to accomplish the net effect of hundreds upon hundreds of additions in the span of less than a minute. Let's now go back to the problem of finding the combined income of a country of 26,784,000 residents where each earns $17,200:

We must calculate 26,784,000 × 17,200.

Repositioning trailing zeros yields: 26,784 × 17,200,000 = (26,784 × 172) *00000.*

We now use the lattice to calculate 26,784 × 172:

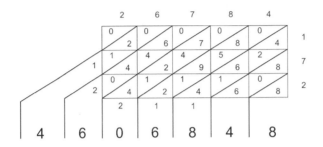

Thus 26,784 × 172 = 4,606,848. Attaching the trailing zeros yields: 4,606,848 *00000.* And now we know that the combined income for all residents in the country is $460,684,800,000.

Using our knowledge about the basic properties of multiplication in HA numerals along with the ability to engage the method on the lattice has enabled us to convert a process which in its raw and original form would easily take more than 300 days to accomplish (using tens of thousands of pieces of paper weighing into the hundreds of pounds) into one which, through an *elegant dance of the digits*, can be accomplished in the space of one-fifth of a single sheet of paper! This is miniaturization on a scale that rivals converting room-sized computers into devices that we can now conveniently sit on our laps. This is conversion on a scale that rivals what we do in language everyday, where, for the purposes of communication, physical events, and abstract thoughts are "road-mapped" into sounds or visible marks—effectively making the inaccessible accessible.

The lattice method was discovered well over a thousand years ago, we believe, possibly in ancient India. The procedure was more widely publicized throughout Europe in early printed arithmetics such as the *Treviso Arithmetic* (1478) and Luca Pacioli's *Summa de Arithmetica* (1494).

The Standard U.S. Algorithm

The lattice method vividly shows how the HA numerals can elegantly solve multiplication problems; however, it is not the only procedure for multiplying those numerals. It is not even the one most commonly taught today in the U.S. curriculum. In his *Summa*, Pacioli demonstrated eight different ways to multiply using HA script. In addition to the lattice method, he included several other methods still in use today. In this section we discuss another of the schemes from Pacioli's text. Given that this algorithm is probably the one most prevalently taught in the United States, we will call it the standard U.S. algorithm, or just the standard algorithm.

The tie that binds all of the methods together is that one must successfully and repeatedly engage the multiplication table to make them work.[2] As we have seen, this requires that we carve up the numbers in such a way that we end up with several partial products that must be added together to get the original or main product.

The standard algorithm can also be obtained from the lattice by simply adding horizontally along the rows as opposed to adding along the diagonals. Explaining it to someone that way, however, requires that the lattice first be developed. In this section, we take a more direct route.

We start with 213×3 and as before we split up the 213 into $(200 + 10 + 3)$:

$$213 \times 3 = 200 \times 3 + 10 \times 3 + 3 \times 3 = 600 + 30 + 9 = 6 \quad 3 \quad 9$$
$$= 2H \times 3 \quad 1T \times 3 \quad 3I \times 3 = 6H \quad 3T \quad 9I = 6H \, 3T \, 9I$$

Let's look at this with the products written vertically:

$$
\begin{array}{ccccccccc}
213 & & 200 & & 10 & & 3 & & 2H & & 1T & & 3I \\
\times\ 3 & = & \times\ 3 & + & \times\ 3 & + & \times\ 3 & = & \times 3 & + & \times 3 & + & \times 3 \\
\hline
& & 600 & + & 30 & + & 9 & & 6H & & 3T & & 9I
\end{array}
$$

Note that for $6\,H\,3\,T\,9\,I$, the coins align perfectly with the place values, which means we can jettison them with no loss of information to obtain 639. The fundamental products used from the times table are $\{2 \times 3, 1 \times 3, 3 \times 3\}$. We can organize this all directly on a single diagram as:

$$
\begin{array}{r}
2\ 1\ 3 \\
\times\quad 3 \\
\hline
6\ 3\ 9
\end{array}
$$

Placing these on a single vertical diagram is straightforward as long as no carries are involved. If carries are required, this complicates things slightly but as with addition if we put the carries into the space above, the situation is cleared up. We illustrate this by looking at 213×4:

$$
\begin{array}{ccccccccc}
213 & & 200 & & 10 & & 3 & & 2H & & 1T & & 3I \\
\times\ 4 & = & \times\ 4 & + & \times\ 4 & + & \times\ 4 & = & \times 4 & + & \times 4 & + & \times 4 \\
\hline
& & 800 & + & 40 & + & 12 & & 8H & & 4T & & 12I
\end{array}
$$

This time the initial alignment is not exact enough to remove the labels. That is, for $8\,H\,4\,T\,12\,I$, we can't simply drop the coin tags since this would give us 8412 which is incorrect. We must first convert 12 *I* into 1 *T* 2 *I* and then carry the 1 *T* to the open space above the tens column:

$$\begin{array}{r} \\ \\ 2\,H \\ \times 4 \\ \hline 8\,H \end{array} + \begin{array}{r} \boxed{1\,T} \\ + \\ 1\,T \\ \times 4 \\ \hline 5\,T \end{array} + \begin{array}{r} \\ \\ 3\,I \\ \times 4 \\ \hline 2\,I \end{array} \xrightarrow{\text{jettisoning the coins}} \begin{array}{r} \\ \\ 2 \\ \times 4 \\ \hline 8 \end{array} \quad \begin{array}{r} 1 \\ + \\ 1 \\ \times 4 \\ \hline 5 \end{array} \quad \begin{array}{r} \\ \\ 3 \\ \times 4 \\ \hline 2 \end{array}$$

Organizing this all directly on a single vertical diagram gives:

$$\begin{array}{r} 1 \\ 213 \\ \times \quad 4 \\ \hline 852 \end{array}$$

The part in the diagram after the coins have been jettisoned (and carries applied) retains a complete memory of the entire process (including the proper location of the place values), so from now on we can simply use it as the jumping off point to obtain answers more quickly. The example of 213 × 9 demonstrates this (remember we multiply from right to left):

$$\begin{array}{r} \\ \\ 213 \\ \times \quad 9 \\ \hline \end{array} = \begin{array}{r} \\ \\ 2 \\ \times 9 \\ \hline 18 \end{array} \quad \begin{array}{r} \\ \\ 1 \\ \times 9 \\ \hline 9 \end{array} \quad \begin{array}{r} \\ \\ 3 \\ \times 9 \\ \hline 27 \end{array} = \begin{array}{r} 1 \\ + \\ 2 \\ \times 9 \\ \hline 19 \end{array} \quad \begin{array}{r} 2 \\ + \\ 1 \\ \times 9 \\ \hline 1 \end{array} \quad \begin{array}{r} \\ \\ 3 \\ \times 9 \\ \hline 7 \end{array}$$

On a single vertical diagram this becomes:

$$\begin{array}{r} 12 \\ 213 \\ \times \quad 9 \\ \hline 1917 \end{array}$$

The fundamental products from the times table this time are 2 × 9, 1 × 9 and 3 × 9.

All multiplications involving any number times a single digit may be performed in the same manner as these examples. How do we extend it to multiplications involving multiple digits? As before, we carve up the multiplication into components involving only single digits and possibly trailing zeros. Let's see how it all plays out in the new format by multiplying 213 × 49:

$$
\begin{array}{c}
213 \\
\times\ 49 \\
\hline
\end{array}
\ =\
\begin{array}{c}
213 \\
\times\ 9 \\
\hline
\end{array}
\ +\
\begin{array}{c}
213 \\
\times\ 40 \\
\hline
\end{array}
\quad
\begin{array}{c}
\text{Using the results above} \\
\ =\
\end{array}
\quad
\begin{array}{c}
1\ 2 \\
213 \\
\times\ 9 \\
\hline
1917
\end{array}
\ +\
\begin{array}{c}
1 \\
213 \\
\times\ 40 \\
\hline
852\ 0
\end{array}
$$

If we want to put this on a single diagram, we can simply stack the rows. In the United States, more often than not, the convention is to stack the rows in ascending order by the number of trailing zeros (i.e., the row with no initial trailing zero goes on top of the row with one initial trailing zero and so on). Doing so for this multiplication yields:

$$
\begin{array}{c}
213 \\
\times\ 49 \\
\hline
1917 \\
8520 \\
\end{array}
\qquad
\xrightarrow{\text{Adding the two rows gives}}
\qquad
\begin{array}{c}
213 \\
\times\ 49 \\
\hline
1917 \quad \longleftarrow \text{No initial trailing zero}\\
8520 \quad \longleftarrow \text{One initial trailing zero}\\
\hline
10437
\end{array}
$$

Which gives $213 \times 49 = 10437$.

In this diagram, where we have combined both rows, we have chosen to leave off the carries. This is done in practice if one simply handles the carries mentally as one proceeds through the multiplication. Of course, the carries can also be written down as well but this can get a bit messy when both rows involve carries as in the previous diagram.

Finally, let's put this all together by seeing how it plays out with our signature multiplication of 745×289:

$$
\begin{array}{c}
745 \\
\times 289 \\
\hline
\end{array}
\ =\
\begin{array}{c}
745 \\
\times\ 9 \\
\hline
\end{array}
\ +\
\begin{array}{c}
745 \\
\times\ 80 \\
\hline
\end{array}
\ +\
\begin{array}{c}
745 \\
\times 200 \\
\hline
\end{array}
\ =\
\begin{array}{c}
4\ 4 \\
745 \\
\times\ 9 \\
\hline
6\ 7\ 0\ 5
\end{array}
\ +\
\begin{array}{c}
3\ 4 \\
745 \\
\times\ 80 \\
\hline
5960\ 0
\end{array}
\ +\
\begin{array}{c}
1 \\
745 \\
\times\ 200 \\
\hline
149\,0\ 00
\end{array}
$$

Stacking the rows in the ascending order of trailing zeros and leaving off the carries gives:

$$
\begin{array}{c}
745 \\
\times 289 \\
\hline
6\ 7\ 0\ 5 \quad \longleftarrow \text{No initial trailing zero}\\
5\ 9\ 6\ 0\ \boxed{0} \quad \longleftarrow \text{One initial trailing zero}\\
1\ 4\ 9\ 0\ \boxed{0\ 0} \quad \longleftarrow \text{Two initial trailing zeros}\\
\hline
2\ 1\ 5\ 3\ 0\ 5
\end{array}
$$

There is nothing that prevents us from stacking the trailing zeros in descending order and doing so gives a slightly different method also employed in the school classrooms of today:

```
              7 4 5
          x 2 8 9
    1 4 9 0 [0 0]   ⟵──── Two initial trailing zeros
        5 9 6 0 [0]   ⟵──── One initial trailing zero
            6 7 0 5   ⟵──── No initial trailing zero
    2 1 5 3 0 5
```

Looking back at the partial sums for 745 × 289, if we choose to add the rows horizontally, we obtain the stacked rows in the standard algorithm (as shown earlier):

$$745 \times 289 \; = \begin{array}{rcccccl} & 140000 & + & 8000 & + & 1000 & \rightarrow \\ + & 56000 & + & 3200 & + & 400 & \rightarrow \\ + & 6300 & + & 360 & + & 45 & \rightarrow \end{array} \qquad \begin{array}{cr} & 149000 \\ + & 59600 \\ + & 6705 \end{array}$$

If we choose to add the values along the diagonals (going from right to left), then we are well on the trail to the lattice method:

$$745 \times 289 \; = \qquad 140000 + 8000 + 1000$$
$$+ \quad 56000 + 3200 + 400$$
$$+ \quad 6300 + 360 + 45$$

Of the two methods developed (lattice or standard), which do you prefer? Most will undoubtedly stick with the one that is most familiar. How do the techniques compare?

The lattice method differs from the standard method in that it does all of the multiplications right up front and then performs the additions. The standard method reserves most of the additions until the end but mixes in some additions with the multiplications when carries are needed.

The lattice method also writes out explicitly everything that is done, including the carries. This has the advantage of making transparent every step involved in the calculation. The standard method is not as transparent—meaning that more memorization is often needed and more effort may be required in retracing one's work.

A complaint often leveled at the lattice method is the lattice itself. For those used to the standard U.S. method, having to write out the lattice can seem like an overly ornate way to multiply simple numbers. In fact, it is probably the difficulty in reproducing the lattice with printing presses that caused the method to fall somewhat out of favor.

Whatever your preference, both the lattice and standard algorithms work spectacularly well in circumventing the tedious process of repeated addition. And showing this fact was a primary goal in these two chapters.

Conclusion

The Hindu-Arabic numerals have spoken! It is now possible to reduce all multiplications of whole numbers, no matter their size, to the one hundred entries contained in the multiplication table. The spirit of this idea is still alive and well to this very day.

One of the major goals of many mathematicians and scientists is to find or classify all of the fundamental patterns in a given area. The hope is then to be able to reduce the full complexity of all behaviors in the given domain to some combination of behaviors involving only these fundamental patterns. Our use of the multiplication table in the algorithms discussed in this chapter provides a vivid illustration of the potential power of this idea.

Our ancestors deserve tremendous credit and recognition for their insights. Their collective efforts have gifted to us ways of multiplying in writing that can in principle be taught to nearly everyone.

How to summarize all of this? There are many avenues to take but we will focus on just one—the power of symbolic maneuver. And while a fair portion of this book is about precisely this dynamic, maneuver is on such glowing display in our discussion of multiplication that it would be almost a shame not to briefly take special notice of it here.

Our discussion began by launching into the physical everyday phenomena of repetitive acts and trying to count them. Our methods in writing, knowing how to only add and subtract with HA numerals, were not up to the task. To remedy this necessitated our taking an extended tour through a world of symbols looking for patterns. We found many, and ultimately employed them in schemes that allowed us to transform massive and overwhelming repetitive tasks into stylishly, swift maneuvers on small diagrams—all the while losing no essential quantitative information.

It is almost as if we took the essence of the situations we encountered and dissolved them into a symbolic cauldron, refashioning them into more potent tools much as we refashion metals into strong and effective tools by heating or melting them. Having done so, it now becomes possible to continually reuse these newly sculpted symbolic tools to more effectively handle the universal problems involving the counting of repetitive acts.

This process is allied with the one of directly applying mathematics to solve physical world problems, but it is not quite the same thing. One aspect of using mathematics to solve real world problems is to transform their essence into mathematical symbols and then apply well-established procedures involving those symbols to obtain an answer which is then translated back to the situation at hand. We do this, for example, every time we perform the simple addition of two dollar amounts at the bank or grocery store.

What has happened here in our discussion on multiplication is that we translated the core of real world problems into symbols, but found the methods there (involving repeated addition) to be inadequate in solving these problems in realistic amounts of time—meaning the traditional route of directly applying mathematics hit a roadblock at this point. This led to an expedition into the world of mathematics itself looking for better symbolic methods (involving HA numerals in this case) to unblock the route. We found them; thus opening the way to systematically handling the multitude of basic problems that involve the counting of repetitive acts.

On the surface, this appears to be nothing different than what we have done earlier in devising our coin system of numerals, the HA numerals themselves and the methods for addition and subtraction. While that is true in principle, it is not true in the details? What we have done in the last two chapters, with multiplication, is a bit more sophisticated symbolically than nearly everything else about numeration that we have previously discussed in this text.

In a sense, everything else we did involving symbols didn't stray too far from the physical problems at hand. That is, even though we were working with symbols (taking symbolic tours, if you will), the physical things we were trying to describe were never far from view. The symbolic tours we have taken in chapters 6 and 7, however, were deeper forays into the mathematical world, where in a sense we momentarily lost actual sight of the physical things we were describing.

This is illustrated by the fact that the HA methods for multiplication, while learnable by the majority who study them, are not initially obvious or intuitive. What this shows, once more, is that exploring the symbolically rich world of mathematics in greater depth (as an entity in its own right), straying far from the physical and concrete, is no trivial matter. It holds the promise not only of yielding greater insight into mathematics itself but also of yielding greater understanding in real world scenarios as well.

What continually impresses and often astounds many mathematicians and scientists is that these explorations can penetrate so deep into mathematics that they stay out of effective sight of physical applications for decades, and then suddenly like a bolt from the blue turn out to be precisely what is needed to solve some modern problem about atoms, gravity, or computers.

This supports our sustained contention that mathematical objects and procedures have much in common with language words and statements—which continually find new uses hundreds of years after their initial creation. According to the Oxford Dictionary, the word "network" has been in use at least since the mid-1500s, making an early appearance in the Geneva Bible of 1560.[3] Whoever first used the word certainly did not have in mind its present-day uses. Yet the word turns out to be precisely what is needed nowadays, turning up all over the place in areas such as computer networks, transportation networks, communication networks, television networks, support networks, and so on. In fact, a 2010 Google search of the word produced 884 million items—more items than either of the words "football" or "sex."[4]

And the eternal relevance and reusability of statements is attested to by the continued popular use of quotations by speechmakers today as well as in the high volume traffic of quotations websites on the Internet.

Enough summary, time to move on. In the next chapter, we explore new viewpoints and find that a whole new and fresh way of reckoning, complete with its own set of issues, will be thrust upon us.

8

The Highest Mathematical Faculties

<hr>

Before the introduction of the (Hindu) Arabic notation, multiplication was difficult, and the division of integers called into play the highest mathematical faculties.

—Alfred North Whitehead (paraphrased), British mathematician,
logician, philosopher and educator[1]

A SYMBOLIC INDUSTRY HAS BEEN bequeathed to us! In simply developing the apparatus for the representation of quantity, we can, with the greatest ease, describe a veritable host of diverse quantities ranging from the population of the world to the gross domestic product of South America to the average depth of the Pacific to the sizes of our kitchens. Now that recipes for adding, subtracting, and multiplying numbers have been brought to the table, we also have convenient access to all sorts of information that previously would have been tedious, if not impossible, for us to obtain. For instance, the problem of determining your entire lifetime contributions to a savings account, if you put away $175 per month for 35 years (requiring that we multiply $175 \times 12 \times 35$), can be answered in a minute or two using the pencil and paper algorithms developed in the last chapter.

Are you ready for another journey? The inverse of repeated addition—the notion of repeated subtraction—awaits us. It is a notion that is even harder to grapple with than its predecessor. But the HA numerals are more than up to the challenge—yielding yet again an effective method to successfully treat these problems as well.

But unlike addition, subtraction, and multiplication, the recipes developed here will not be capable of completely eliminating trial and error from the process. Solving the problems of repeatedly taking away objects was considered hard even by medieval abacists, some of whom likened the difficulty to the hardness of iron. Even the modern day algorithms for treating these problems remain the source of constant and intense debate among educators. It is probably safe to say that a supermajority of those who have learned the processes for dealing with repeated subtraction still don't know why the methods really work.

In the next two chapters, we head full bore into this world. Will our conceptual tools be up to the task of explaining the processes involved? We shall put them to the test and find out. As always, you, the reader, will be the final arbiter on their effectiveness for you.

Repeated Subtraction

We have previously considered such questions as: How many crates of coffee are shipped to a coffee shop in five months, if every shipment consists of seven crates. Given that this process builds to a collection of thirty-five total crates, a natural question the supplier of the crates might ask in reverse (of the coffee shop owner) would be: If we have thirty-five crates to start with, how many times can we take away crates, seven at a time, until the collection is exhausted? We will find that this latter question is generally a harder one to answer.

Let's answer this question directly by performing the required subtractions. For the sake of convenience, we will symbolically represent the thirty-five crates by thirty-five circles:

⊙⊙⊙⊙⊙⊙⊙⊙⊙⊙ ⊙⊙⊙⊙⊙⊙⊙⊙⊙⊙ ⊙⊙⊙⊙⊙⊙⊙⊙⊙⊙ ⊙⊙⊙⊙⊙	− ⊙⊙⊙ ⊙⊙⊙⊙	=	⊙⊙⊙⊙⊙⊙⊙⊙⊙ ⊙⊙⊙⊙⊙⊙⊙⊙⊙ ⊙⊙⊙⊙⊙⊙⊙⊙⊙ 1st Subtraction
⊙⊙⊙⊙⊙⊙⊙⊙⊙⊙ ⊙⊙⊙⊙⊙⊙⊙⊙⊙⊙ ⊙⊙⊙⊙⊙⊙⊙	− ⊙⊙⊙ ⊙⊙⊙⊙	=	⊙⊙⊙⊙⊙⊙⊙⊙⊙ ⊙⊙⊙⊙⊙⊙⊙⊙⊙ ⊙ 2nd Subtraction
⊙⊙⊙⊙⊙⊙⊙⊙⊙⊙ ⊙⊙⊙⊙⊙⊙⊙⊙⊙⊙ ⊙	− ⊙⊙⊙ ⊙⊙⊙⊙	=	⊙⊙⊙⊙⊙⊙⊙⊙⊙ ⊙⊙⊙⊙ 3rd Subtraction
⊙⊙⊙⊙⊙⊙⊙⊙⊙⊙ ⊙⊙⊙⊙	− ⊙⊙⊙ ⊙⊙⊙⊙	=	⊙⊙⊙⊙⊙⊙⊙ 4th Subtraction
⊙⊙⊙⊙⊙⊙⊙	− ⊙⊙⊙ ⊙⊙⊙⊙	=	5th Subtraction

We see that five subtractions are required to exactly exhaust the collection.

Rephrasing this question slightly differently gives us the ability to answer the question with much less writing. Instead of asking, how many times can we take seven crates away from thirty-five until the collection is exhausted, we may ask how many groups of seven crates can we form out of thirty-five. In this mind-set, we rearrange the thirty-five circles into groups of seven:

Grouping in sevens yields

We see that there are five groups of seven circles. Each subtraction of seven circles takes one of these groups away from the collection. Consequently, this repackaging of the 35 circles makes transparent the fact that we must perform five subtractions to exhaust the set—meaning that we can tell how many times we have to subtract without actually doing the subtractions.

Let's now use this latter technique to see how many times we can take five objects away from forty until the collection is exhausted.

We first rephrase the question as: How many groups of five objects can we form out of forty?

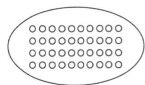 Grouping in fives yields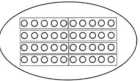

There are eight groups, hence our answer to both questions is 8. We summarize the two examples in the following table:

Summary of Repeated Subtractions of 35 and 40 Objects

Question	Size of Collection	*Size of Groups*	*# of Groups*	Answer to Question
How many groups of 7 crates can we form from thirty-five?	35	7	5	5 groups
How many groups of 5 objects can we form from forty?	40	5	8	8 groups

As usual, our ultimate goal is to streamline the process of repeated subtraction so that we don't have to rely on clumsy diagrams. The above table gives a clue on how to proceed down this path. Clearly exhibited is the general property that:

the size of the collection = (size of the group) × (the number of groups)

Let's enlist the help of a table to answer the question of how many groups of 9 objects can we form out of 63?

Size of Collection	Size of Groups	# of Groups
63	9	??

To obtain the answer we simply need to find the number which when multiplied to 9 yields 63. This number we know must be 7. Thus we see that the method of using the table and multiplication is much simpler than the brute-force method of carving out groups of nine circles from a collection of sixty-three.

Now that we have rerouted the problem of repeated subtraction into a tabular form, let's play with it a bit. We have been given the size of the collection and asked to find the number of groups of that size we can carve the collection into. Let's now swap the question and consider the scenario where we know the number of groups instead and want to find the size of each group. In this case the last question reads as: If we carve up 63 objects into 7 equally sized groups, what will be the size of each group? Using the table, we have:

Size of Collection	Size of Groups	# of Groups
63	??	7

We know that since the "(size of the groups) × 7" should be 63, our answer will be 9 objects per group.

Now consider the down to earth scenario involving 66 international students at a small college. If they are placed into six classes of equal size, what will be the size of each class? The table sets up as:

Size of Collection (# of Students)	Size of Groups (# of students in each class)	# of Groups (# of Classes)
66	??	6

We must have: "(size of the groups/classes) × 6 = 66" which gives 11. Thus each class must have 11 students.

A Three-Sided Coin

Taking stock of our progress to this point, we see, in essence, three types of problems present in our work. And though phrased differently, each problem can be solved in the same way:

Type A. Given a collection of objects, how many times can we take away a specific number of objects from the collection until it is exhausted?

Type B. Given a collection of objects, if we organize the objects of the collection into groups of a certain size, how many groups will there be?

Type C. Given a collection of objects, if we distribute the objects equally among a certain number of groups, what will be the size of each group?

Let's now observe how each of these questions can be played out on the same collection.

A: Given a collection of 24 objects, how many times can we take away 8 objects from the collection until it is exhausted.

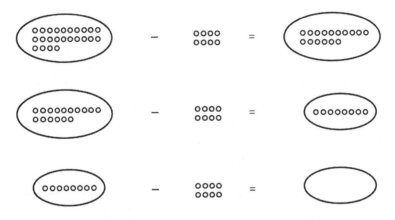

The answer is 3 times.

B: Given a collection of 24 objects, if we organize the objects of the collection into groups each containing 8 objects, how many groups will there be?

The answer is 3 groups.

C: Given a collection of 24 objects, if we partition or distribute them equally among 8 groups, what will be the number of objects in each group?

```
OOOOOOOO
OOOOOOOO
OOOOOOOO
```
24 objects   1   2   3   4   5   6   7   8

8 groups

Distributing objects
equally yields
```
O   O   O   O   O   O   O   O
O   O   O   O   O   O   O   O
O   O   O   O   O   O   O   O
1   2   3   4   5   6   7   8
```

The answer is that every group will have 3 objects.
How are A, B, and C are related?

A and B: All we need do is perform B first and then subtract. Clearly once we organize the objects into the groups of 8, we can simply count the number of groups or we can repeatedly subtract them; in either case, the answers will be equal.

B and C: Imagine that the 24 objects are cards to be dealt among 8 players. In such a scenario, each subtraction of 8 objects corresponds to a deal of a single card to every player.

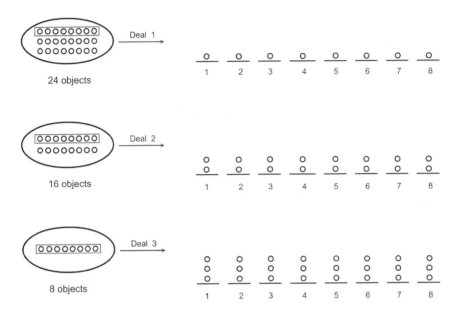

There are three subtractions of 8 objects and thus 3 deals, hence each group consists of 3 objects or each player receives three cards. Thus if we interpret subtractions as deals, in this context, we see that the solutions to B and C are related.

We see that A and B correspond to viewing the 24 objects as 3 groups of 8 objects. Whereas, C corresponds to viewing those same 24 objects as being partitioned into 8 groups, each of which contains 3 objects.[2] A geometrical way to phrase it is that we can view twenty-four objects as consisting of 3 rows of 8 objects each or as 8 columns of 3 objects each.

There is yet again an infinite nation of scenarios which correspond to Type A, Type B, or Type C problems. The next example illustrates these problems at play in a real world context.

You owe $750 to a friend. She allows you to make monthly payments of $50. How many months will it take you to pay off the debt?

This is phrased in the context of problem types A and B. Each payment subtracts $50 from the $750 owed, so the debt will be paid off when $50 is subtracted enough times to exhaust $750. We can also think of organizing the $750 into $50 packets and count the total number of packets. In this case, each packet is the symbolic representative of one month.

Using multiplication to help, we have that "(the number of months) × 50 = 750." The number of months to make this equation true is 15. Hence in 15 months the debt will be paid-off.

This scenario can also be thought of in the context of problem Type C, but for this problem, thinking of it in that manner seems less natural.

Let's now consider the related problem where the number of months is known but the amount of each payment is not. This time the problem reads as:

You owe $750 to a friend. She allows you a total of 15 months to pay off the debt. If you make payments in equal amounts, how much do you owe each month?

This is phrased in the context of problem Type C. We need to distribute the $750 equally over 15 months. Using multiplication to help again, we have that: (the amount of each payment) × 15 = 750. The amount that makes this equation true is 50. Hence, we must pay $50 each month.

We could also answer this question by asking how many times can we subtract 15 from 750, in the context of problem Type A, but for this scenario it doesn't seem natural to subtract 15 months repeatedly from $750.

To nail down the distinction between the different processes involved in problem Types A/B and C, let's consider a few more scenarios.

A rental car company bought 8 cars of a certain model from General Motors. The total bill was $120,000. If they paid the same price for each car, what was the cost per car to the company?

This is phrased in the context of problem Type C. We spread the $120,000 equally over the 8 cars. To find the answer we need to find the number such that: (the number) × 8 = 120,000. The number which works is 15,000. Thus each car costs GM $15,000.

A construction company can build a home in 35 days. How many homes can they build in 700 days at the same rate?

This is phrased in the context of problem Types A and B. We effectively want to know how many packets of 35 days can be formed out of 700 days. Each packet corresponds to a built home, hence, a determination of this number will give us what we desire. We need to find the number such that: (the number) × 35 = 700. This number is 20. Thus the company can build 20 homes in 700 days.

A New Way of Reckoning

It is interesting to note that, although we can interpret these problems in different ways, the same method can be used to solve them. It is this common method of solution that we now wish to study in some detail. The first item of business is to give a name to the method and elevate it in short order to the status of a genuine operation on numbers. Today we use the word "division" to describe this way of reckoning.

Thus if we want to know how many groups we will have from 40 objects arranged into groups of size 8, we will say that the answer can be found by calculating: 40 divided by 8. We represent the method by using special symbols ("÷" and "—") like so:

I. $40 \div 8$

II. $\frac{40}{8}$

Problems requiring division, like those involving addition, subtraction, multiplication, date back to antiquity. The Babylonians, like us, evidently used a special symbol for division while the Greeks seemed to have often used their language words instead (the equivalent in English would read something like 40 *divided by* 8 as opposed to using either of the special symbols in I or II). Evidently in ancient India, the number being divided into was sometimes written on top of the number doing the dividing with no line between them (e.g., $\frac{40}{8}$). It is thought that the fraction bar was first used by the Arabs in the 1100s.[3]

Math historian Florian Cajori claims that the symbol (÷), called an obelus, was first introduced into the literature for representing division in 1659, by the Swiss mathematician, Johan Heinrich Rahn.[4] The Merriam-Webster dictionary defines the obelus as: "a symbol – or ÷ used in ancient manuscripts to mark a questionable passage."[5] This practice, dating at least back to the 200s BCE, of using either of these two symbols in the same manner, perhaps gives a partial reason as to why the symbol (÷) was also sometimes used to represent subtraction. The symbol, as division, eventually came into regular use in the United States and the British Commonwealth but not so throughout many parts of the world, where the colon (:), introduced by Gottfried Leibniz, is still often used to represent division.[6]

Now that we know that problem Types A, B, and C are equivalent in their method of solution, we can reciprocally interpret division in any of these ways. Hence, 40 divided by 5 or $\frac{40}{5}$ can be interpreted as answering any of the following three questions:

Type A. How many times can we take away 5 objects from a collection of 40 until the collection is exhausted?

Type B. How many groups of 5 can we organize out of 40?

Type C. If we organize 40 objects into 5 equal portions, what will be the size of each of the portions?

We can, of course, answer any of these by finding out "(what number) times 5 is equal to 40"? The result is 8: therefore $\frac{40}{5} = 8$.

In a similar fashion we can answer, $\frac{110}{11} = ?$, by finding out what number times 11 gives 110. This number is 10.

The operation of division represents a unification of different perspectives from the standpoint of how they read in English. In Robin Hood–like fashion, it ties together the apparently dissimilar things of "repeatedly taking away objects" and "sharing them equally." And even though we can now think of division and multiplication as inverse processes (allowing us to use multiplication as an aid in the solution of division problems), division opens up vast new vistas by exposing us to situations for which there is no counterpart in multiplication.

Division by Zero

In elementary arithmetic, we may represent what we really care about both with symbols, unrelated to English, and with names in English. Having done so, however, doesn't mean that we actually have complete control over everything that we are talking about. Just as the construction engineer controls much in the way that infrastructures are built but still must have an abiding respect for the laws of nature, so too must the mathematician have a healthy regard for the principles of logic.

When we incorporate into the discussion our newly discovered number zero with the operation of division, we find ourselves in a dilemma. In a word, the properties of zero and the previously discussed properties of division just don't mix that well.

Let's consider the specific case of "$\frac{6}{0}$"? In the context of repeated subtraction, this reads as: How many times can we subtract 0 objects from 6 until the collection is exhausted? It is impossible to exhaust a collection in this way, since taking no objects away from a collection does nothing to reduce the size of the collection.

Interpreting it in the context of equal distribution is of no help either. Thinking of division this way the question reads as: If we organize 6 objects into 0 equal portions, what will be the size of each of the portions? In this form, the question doesn't even make sense.

Given that we can't make sense of these foundational interpretations when we divide by zero, should we simply outlaw it? Maybe, but we need more justification than we have so far. Nowadays, in mathematics, we don't ban an idea just because it doesn't make sense when read a certain way. Throughout history, mathematicians have chosen to sometimes act in this fashion much to their later collective regret. The examples of negative numbers, irrational numbers, "imaginary numbers," and even the idea of zero itself stand as eternal testaments, in the historical case file, to the hazards of this practice.

When faced with such predicaments in the past, what has often worked is that some mathematicians forged ahead anyway, using the internal logic of the symbols to navigate through the confusion (hoping that useful interpretations could later be found)—much like a pilot uses instruments such as radar to guide a plane through the fog to a safe landing. Remember, when the symbols are allowed to really fly they can (and most often do) take us places that we can't otherwise go.

Let's see what working with symbols and operations can do for us here. In the case of trying to exhaust a collection of 6 objects by repeatedly taking away zero objects, working symbolically is still of no use. For example, if we subtract zero four times from six, we get nowhere

$$(\text{e.g.,} 6 - \underbrace{0 - 0 - 0 - 0}_{4 \ times} = 6)$$

No matter how many 0s we subtract symbolically, the 6 still remains.

What about enlisting the aid of multiplication? This idea fails as well since to solve "$\frac{6}{0} = ?$" , would mean that we need to find a number such that, $0 \times$ (the number) = 6. This is impossible since any number multiplied by 0 will equal zero not 6.

We encounter the same issues if we replace the six with 8, 20, or any other number. Based on all that we have done above, dividing by zero just doesn't seem to work—at least when the number on top is not zero. We simply can't make a go of it without contradicting fundamental and observable truths in multiplication—truths such as (any number) multiplied to zero gives zero (e.g., $6 \times \big| = \big|$ in abacus rods which translates to $6 \times 0 = 0$). And we are not going there.

While, there appears to be no way of salvaging the case of a nonzero number divided by zero, will the same be true if we replace the nonzero number on top by zero. In other words, do we encounter serious obstacles if we divide zero by zero (i.e., $: \frac{0}{0}$)?

In the context of repeated subtraction this reads as: How many times can we subtract 0 objects from 0 until the collection is exhausted? This time instead of being too restrictive the rules are too generous. Symbolically, we have:

Subtractions of Zero	Number of Times
0	0 times
$0 - 0 = 0$	1 time
$0 - 0 - 0 = 0$	2 times
$0 - 0 - 0 - 0 - 0 = 0$	4 times

Here we see that any of the numbers {0, 1, 2, 4} work according to our definition of division. In fact, we could subtract zero any number of times and still get the same result. According to our earlier definitions of division, zero divided by zero can be any of these numbers.

If we enlist multiplication as an aid, among the possibilities we have the following:

I. Since (the number *0*) $\times$ 0 = 0, we could claim that $\frac{0}{0}$ = 0.

II. Since (the number *1*) $\times$ 0 = 0, we could claim that $\frac{0}{0}$ = 1.

III. Since (the number *2*) $\times$ 0 = 0, we could claim that $\frac{0}{0}$ = 2.

IV. Since (the number *4*) $\times$ 0 = 0, we could claim that $\frac{0}{0}$ = 4.

Any whole number will work here as well according to our established ways of thinking about division. Believe it or not, having every number work is as problematic as having no number that works.

Imagine a database of records. Databases generally are set up such that every record has an identification number, often called a primary key. In many databases, items such as a Social Security number, credit card number, or phone number are used as the identifier. Consider the following two situations for a database with 100,000 records: (1) No identification number is given to any of the records and (2) Every record is given the same identification number. Neither of these is desirable since we cannot easily identify or search records in either scenario. For all intents and purposes, a database created in either of these fashions is ineffectual.

The first situation is roughly analogous to the case of $\frac{\text{``\textit{a nonzero number}''}}{0}$, where no number will work and the second situation is roughly analogous to $\frac{0}{0}$, where every number works.

Worse yet, allowing $\frac{0}{0}$ to have meaning will cause paradoxes to occur. For example, let's say we decided to make a choice and let $\frac{0}{0}$ = 1. This simple choice could lead to catastrophe. Consider the case in multiplication where we have 5 $\times$ 4 $\times$ 2 = 20 $\times$ 2. We can divide both sides of the equation by 2 like so: $\frac{5 \times 4 \times 2}{2} = \frac{20 \times 2}{2}$. Then canceling out the 2s and replacing them by 1 (we can do this because $\frac{2}{2}$ = 1) yields 5 $\times$ 4 = 20. We could more easily do the division mentally and just cancel out the 2s directly like: 5 $\times$ 4 $\times$ 2̸ = 20 $\times$ 2̸, to obtain 5 $\times$ 4 = 20.

Now let's see how this process breaks down if we cancel out or divide by zero the way we did with the 2. Since any expression times zero is equal to zero we know that 5 $\times$ 4 $\times$ 0 = 155 $\times$ 0 = 0. Now if we cancel out the 0s the way we did the 2s (this would be possible if we have $\frac{0}{0}$ = 1), we obtain 5 $\times$ 4 $\times$ 0̸ = 155 $\times$ 0̸.

Removing the zeros implies that $5 \times 4 = 155$, which we know to be nonsense. In fact using this same reasoning, we could show that 5×4 equals any other number (simply replace 155 by the value of your choice). Similar paradoxes occur if we allow $\frac{0}{0}$ to be any other number as well. This is simply unacceptable.

Moreover, this situation is a transparent contradiction since only a couple of steps are involved. In cases involving more intricate reasoning (such as occurs in algebra and beyond), if we were to allow division by zero, we could generate all sorts of contradictions and not know from whence they sprang. This often happens when trick problems involving long algebraic arguments are given. Often in such cases the final answer seems to imply a contradiction such as $1 = 2$. More times than not, the likely perpetrator is a division by zero which has been disguised. Mathematicians are simply left with no choice except to outlaw division by zero.

Zero strikes again! It first shocked us by its very arrival, announcing to the world that there exist numbers out there that are masked from normal view and whose existence may only be revealed to us through symbolic manipulations in mathematics. Now it astonishes us yet again by showing that it refuses to completely submit to division. Its influence throughout the breadth and depth of mathematics is no less spectacular. The simple idea of the empty rod on the abacus has traveled very, very far indeed.

Remainders

Division also differs from multiplication in another crucial respect. In multiplication, the result of the physical process of repeatedly adding a whole number quantity, say for instance an $8 payment, can always be described by a whole number. That is, if we wanted to repeatedly add one hundred payments each $8 strong, we obtain a dollar amount described by the single whole number 800. In division this does not always occur as we observe in the next example.

Consider $14 \div 4$. We count the number of times that we can subtract 4 objects from 14, until the collection is exhausted. We can easily subtract 3 groups of 4 objects from 14 objects; however, the collection is not exhausted (2 objects are left over), as shown here:

How are we to describe the result of this process of dividing 14 into groups of 4? We certainly cannot represent it as 3, for $\frac{14}{4} \overset{?}{=} 3$, would imply by previous reasoning that $4 \times 3 = 14$ which is not true. We cannot represent it by 4 either, since we can only subtract 4 at most 3 times from 14. Since 3 and 4 are the only whole numbers that have a chance, it would appear that no single whole number can describe the result of this process. This yields yet another dilemma. Since there is no single whole number to describe it, is this a division that we should outlaw as we did in the case of division by zero?

Not so fast. This situation is different from the case of division by zero. Here we can at least depict the result of this process in an intuitive manner that is systematic and reproducible. We can describe $14 \div 4$ as yielding 3 groups of 4 objects with 2 objects left out—there is certainly nothing strange or bizarre about that. We can even create a notation to indicate this by saying that $14 \div 4$ is equal to 3 with a remainder of 2—abbreviated to 3 R 2. Division by zero allowed no possible way out of the dilemma without drastically altering "sacred truths."

Let's consider $18 \div 7$. We can organize 18 into two groups of 7 with 4 objects left over. Thus this result or quotient can be described by 2 R 4:

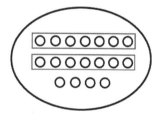

We can work with remainder notation in other ways to find out useful information as well. Consider the scenario where we know that organizing a collection of fighter jets into squadrons of 12 planes yields the result 5 R 4. Can we find the total number of planes?

Stated another way, we know that the process of organizing the jets into twelves gives 5 such squadrons with four planes left out. We easily find the total number of jets by simply calculating $5 \times 12 + 4$. This gives a collection of 64 planes.

A single whole number cannot describe a situation where the organization of a collection into groups of a certain size leaves some objects out—there are two numbers to keep track of, the number of groups and the number of objects left out. The introduction of remainder notation (involving 2 whole numbers in combination) gives us the ability to describe these new scenarios. It is, however, a new notation and questions as to what these symbolic entities actually represent must be asked. Whole numbers can stand on their own; by which it is meant that the number represented by the symbol 3 has a meaning independent of the fact that 12 divided by 4 yields 3.

Can something like 5 R 4 stand on its own? Is it the representation of a true number or simply a convenience helping us with certain division problems? If entities such as 5 R 4 do represent actual numbers, it would be natural to inquire about their arithmetic; how would we add, subtract, multiply, and divide them? This puts us in the arena of fractions which are not specifically discussed in this text. At present, we will only think of remainder notation as a convenience for describing certain types of situations occurring in division. When the remainder is zero, the notation is not needed and we continue as before, simply using the whole number alone.

Peering into Division

So far we have only considered simple division problems—problems whose answers can be obtained almost instantaneously through the use of multiplication. For more difficult divisions, such as $\frac{63438}{654}$, an instantaneous solution is rarely possible. This means that we will need to attempt to develop systematic techniques that allow us to circumvent both the method using clunky diagrams and to some extent the trial-and-error method using multiplication (i.e., guessing in one grand gesture what number multiplied to 654 equals 63,438). Before discussing the standard long division algorithm that gives a systematic technique (while not entirely limiting the guesswork), we first consider two methods that will allow us to simplify certain types of division problems. The insights gained from looking at these will prove useful later when discussing long division.

Shortcuts Using Multiplication and Trailing Zeros

The free agent properties of trailing zeros used so effectively in multiplication can also be exploited in division:

$\frac{8}{2} = 4$ since $2 \times 4 = 8$; $\frac{80}{2} = 40$ since $2 \times 40 = 80$

$\frac{800}{2} = 400$ since $2 \times 400 = 800$; $\frac{8000}{2} = 4000$ since $2 \times 4000 = 8000$

Observe that all we need do is divide 8 by 2 to obtain 4 and then attach the number of trailing zeros. We can extend this idea, so for instance, in the quotient, $\frac{8\,000000}{2}$, the dividend (top number) has 6 zeros. So all we need do is to divide 8 by 2 to obtain 4 and then attach the 6 zeros like so: $\underbrace{8\,000000}_{} = \underbrace{4000000}_{6\,zeros}$. This holds in general.

When trailing zeros are both on the dividend and the divisor (bottom number), we can bring out the cutting knife and cancel zeros in equal measure as shown in the table:

Straight Division	First Canceling Out Trailing Zeros Before Dividing $\rightarrow$	Then Dividing
$\dfrac{60}{30} = 2$	$\dfrac{6\cancel{0}}{3\cancel{0}}$	$\dfrac{6}{3} = 2$
$\dfrac{600}{30} = 20$	$\dfrac{60\cancel{0}}{3\cancel{0}}$	$\dfrac{60}{3} = 20$
$\dfrac{600}{300} = 2$	$\dfrac{6\cancel{0}\cancel{0}}{3\cancel{0}\cancel{0}}$	$\dfrac{6}{3} = 2$
$\dfrac{8000}{400} = 20$	$\dfrac{80\cancel{0}\cancel{0}}{4\cancel{0}\cancel{0}}$	$\dfrac{80}{4} = 20$
$\dfrac{8000}{40} = 200$	$\dfrac{800\cancel{0}}{4\cancel{0}}$	$\dfrac{800}{4} = 200$
$\dfrac{8000}{4000} = 2$	$\dfrac{8\cancel{0}\cancel{0}\cancel{0}}{4\cancel{0}\cancel{0}\cancel{0}}$	$\dfrac{8}{4} = 2$

And you can see that, if we cut out the trailing zeros first and then divide (third column), we obtain the same answer as we do dividing all at once (first column).

We are still far from a general algorithm, but what we can accomplish at this point already can be a huge time-saver. Recall that $\frac{30,000,000}{15,000}$ is a process which may be interpreted in these three ways:

Type A. How many times can we subtract 15,000 from 30,000,000?

Type B. How many groups containing 15,000 objects can we form out of 30,000,000 objects?

Type C. If we divide 30,000,000 into 15,000 equal portions, what will be the size of each of these portions?

With not too much effort, we are able to once again circumvent directly carving up a collection of 30,000,000 objects to ascertain that the answer in all three cases is 2,000. The symbolism has once again revealed patterns that we can take advantage of; moreover, the answer is true regardless of whether the 30,000,000 objects are people, cars, dollars, rocks, or stars. The pieces are completely interchangeable without affecting the way we solve the problem.

Can We Alphabetize Division?

Knowing that it is possible to effectively alphabetize addition, subtraction, and multiplication, it is natural to wonder if the same is true of division. Can we construct a division table in the same manner as the times table, ultimately reducing all complicated divisions to simple ones from a small chart?

It is certainly possible to construct a basic table of sorts for small divisions, but unfortunately, a scheme such as the one employed using the multiplication table is not possible. A major reason for the clean, tidy procedure developed for multiplication is that the distributive property holds in a symmetric fashion; that is, we can calculate 32×26 by first taking apart the 32 and writing "$(30 + 2) \times 26$" as "$30 \times 26 + 2 \times 26$" (after which we then carve up the 26) or we can calculate 32×26 by first breaking up the 26 and writing "$32 \times (20 + 6)$" as "$32 \times 20 + 32 \times 6$" (and then decomposing the 32). We can disassemble both numbers in the product, let the times table run loose on it, and still obtain the correct answer.

The distributive property does not hold for division in this same symmetric fashion. While we can carve up the dividend (left hand or top term) and still obtain correct results, we cannot do the same with the divisor (right hand or bottom term).

We demonstrate both cases here for the division of 24 by 8:

I. Disassembling the left hand or top term (gives correct answer): We know that $24 \div 8 = 3$ or $\frac{24}{8} = 3$.
   Graphically this plays out as:

If we disassemble the 24 and rewrite it as $16 + 8$, then distribute the division by 8 to both numbers we obtain the same answer:

$$24 \div 8 = (16 + 8) \div 8 = 16 \div 8 + 8 \div 8 = 2 + 1 = 3$$

or

$$\frac{16 + 8}{8} = \frac{16}{8} + \frac{8}{8} = 2 + 1 = 3$$

Graphically taking the 24 objects and breaking them into 2 smaller collections of sizes 16 and 8, respectively, and distributing them each over the 8 slots gives:

$16 \div 8$:

16 objects — 2 Deals → Each Portion Size is 2 Circles

$8 \div 8$:

8 objects — 1 Deal → Each Portion Size is 1 Circle

Adding up the portion sizes in both allocations still gives a total of 3 circles.

II. Disassembling the right hand or bottom term (gives a wrong answer in general):

Rewriting the 8 in the divisor as 6 + 2 and then separately distributing the 6 and 2 yields:

$$24 \div 8 = 24 \div (6 + 2) \stackrel{??}{=} 24 \div 6 + 24 \div 2 = 4 + 12 = 16$$

or

$$\frac{24}{6+2} \stackrel{??}{=} \frac{24}{6} + \frac{24}{2} = 4 + 12 = 16$$

The answer of 16 here does not agree with the 3 we obtained in I. There is no doubt that the correct answer for $24 \div 8$ is 3, so we must conclude that something goes awry when we carve up the divisor and distribute it. Let's see if a graphical viewpoint gives us some insight on why this fails:

$24 \div 6$:

24 objects — 4 Deals → Each Portion Size is 4 Circles

$24 \div 2$:

24 objects — 12 Deals → Each Portion Size is 12 Circles

Blindly, adding these 2 portion sizes up gives us 16 portions, but it can readily be seen that the process in II is very different from the processes in I. In II, we have taken one collection of 24 objects and broken it up into 6 equal portions and then have taken another collection of 24 objects and distributed it into 2 equal portions—meaning in effect, that we are working with 48 total objects (unequally distributed in portions of 6 and 2) whereas in I we are only working with 24 total objects equally distributed into the same 8 portions. These processes simply are not equivalent and it should come as no surprise that they yield different results.

What this means is that, unlike in multiplication, we will not be able to break up both numbers involved in a division such as $\frac{63438}{654}$ in a symmetric way. We can still carve up 63438 as $60000 + 3000 + 400 + 30 + 8$ and split up the division accordingly as: $\frac{60000}{654} + \frac{3000}{654} + \frac{400}{654} + \frac{30}{654} + \frac{8}{654}$. But since 654 occurs on the bottom, we can't carve it up as $600 + 50 + 4$ and split up the divisions. That is, $\frac{63438}{654}$ is not equal to $\frac{63438}{600} + \frac{63438}{50} + \frac{63438}{4}$. So, unfortunately, we will not be able to work our magic on the trailing zeros in $600 + 50 + 4$ (as we are able to do in multiplication) but will instead have to take the 3 digits in 654 together all at once.

This will cause the procedure we ultimately develop to involve some measure of guesswork.

Shortcuts by Decomposing the Dividend

Although we can't decompose the divisor (bottom number), we still can use the fact that it is possible to carve up the dividend to help us solve division problems. The example of $\frac{348}{6}$ will help us shed light on this procedure.

We need to rewrite 348 into a sum of numbers each of which we know is divisible by 6. One such partition is: $348 = 300 + 48$. Using this yields:

$$\frac{348}{6} = \frac{300 + 48}{6} = \frac{300}{6} + \frac{48}{6} = 50 + 8 = 58.$$

Thus $\frac{348}{6} = 58$. We verify this by noting that $6 \times 58 = 348$.

We can break up a number into more than two parts if needed to complete a division. Consider this for $\frac{675}{25}$:

$$\frac{675}{25} = \frac{500 + 100 + 75}{25} = \frac{500}{25} + \frac{100}{25} + \frac{75}{25} = 20 + 4 + 3 = 27.$$

This gives $\frac{675}{25} = 27$, which we verify by noting that $25 \times 27 = 675$.

We also can involve subtraction in this game. Let's consider the situation involving determining the monthly payments on a 9-month loan of $864:

$$\frac{864}{9} = \frac{900 - 36}{9} = \frac{900}{9} - \frac{36}{9} = 100 - 4 = 96$$

The payment each month will be $96.

This manner of decomposing the dividend critically depends on what the divisor is. Dividing 4560 by three different numbers gives illustration to this:

A. $\frac{4560}{6} = \frac{4200}{6} + \frac{360}{6} = 700 + 60 = 760$. *Decomposition:* 4560 = 4200 + 360

B. $\frac{4560}{8} = \frac{4000}{8} + \frac{560}{8} = 500 + 70 = 570$. *Decomposition:* 4560 = 4000 + 560

C. $\frac{4560}{12} = \frac{3600}{12} + \frac{960}{12} = 300 + 80 = 380$. *Decomposition:* 4560 = 3600 + 960

These divisions have a whimsical air about them requiring that we know how to break up 4560 into components that are divisible by 6, 8, and 12, respectively. This will not be the case generally (e.g., how do we break 4560 up into components that are each obviously divisible by either 57 or 285).

What we would like to have is a method which will allow us to systematically construct the result of dividing one number by another number regardless of what two numbers are involved. The method of breaking up the dividend does still hold the key. However, instead of breaking up the dividend (top number) into components in individual ways that depend on which number we are dividing by, we will break it up systematically by place values.

A difficulty occurs, however, when we do this. For instance, if we choose to decompose 4560 into components by place values, this will yield 4560 = 4000 + 500 + 60. If we divide this decomposition by 6, 8, or 12, we will, in each case, encounter remainders at some point in the division:

E. $\frac{4560}{6} = \frac{4000}{6} + \frac{500}{6} + \frac{60}{6}$, the first two divisions leave a remainder.

F. $\frac{4560}{8} = \frac{4000}{8} + \frac{500}{8} + \frac{60}{8}$, the last two divisions leave a remainder.

G. $\frac{4560}{12} = \frac{4000}{12} + \frac{500}{12} + \frac{60}{12}$, the first two divisions leave a remainder.

What are we to do about these remainders? If we want to successfully complete divisions using this method, we are going to meet them. This will be true not just for the number 4560 but also in general. Our task in the next chapter will be to seek out a general procedure which will allow us to both systematically break apart numbers by place values while at the same time giving us a way to handle the resulting remainders. While we won't be able to get quite the systematic and alphabetic scenario obtained in multiplication, we still get a nice, workable algorithm.

9

The Powder Keg of Arithmetic Education

It's time to recognize that, for many students, real mathematical power, on the one hand, and facility with multidigit, pencil-and-paper computational algorithms, on the other, are mutually exclusive. In fact, it's time to acknowledge that continuing to teach these skills to our students is not only unnecessary, but counterproductive and downright dangerous.

—Steven Leinwand, contemporary author,
researcher at American Institutes for Research[1]

There is a long standing consensus among those most knowledgeable in mathematics that standard algorithms of arithmetic should be taught to school children. Mathematicians, along with many parents and teachers, recognize the importance of mastering the standard methods of addition, subtraction, multiplication, and division in particular.

—David Klein, contemporary mathematician, math educator and
author and R. James Milgram, contemporary mathematician, author,
former member of the National Board for Education Sciences[2]

THE TIME HAS COME to address what is generally considered one of the most difficult procedures for students to master in all of elementary arithmetic: the long division algorithm (LDA). As the quotations suggest, considerable disagreement (and that is putting it mildly) exists among mathematical educators, parents, research mathematicians, scientists, engineers, and the general public about whether the method should continue to be taught (see the October 2002 issue of *Discover Magazine* for an interesting and civil roundtable discussion on this issue).

Our goal here is not to directly enter this specific fray but rather to focus on magnifying visually the processes that long division abbreviates. The stage has already been set in the last chapter with our discussion on systematically breaking apart the dividend (top number) by place values. We now expand on this notion.

Division with Coin Numerals

So far our illustrations of the processes at play in division have used faceless circles. What will these processes look like if we use coin numerals in their place? Besides simplifying the number of steps required, we will also find that, since coin numerals represent "place values in a box," involving them in the process puts us firmly on the trail to our modern recipe for long division.

We start with a simple example or two and work our way to more interesting cases.

For what follows, it will be more useful to think of division in terms of equal apportionment (Type C of chapter 8). Consider the following two divisions:

$$\frac{6}{3} \text{ and } \frac{600}{3}$$

A. $\frac{6}{3}$

We represent 6 by Ⓘ Ⓘ Ⓘ Ⓘ Ⓘ Ⓘ and 3 by

Distributing the six coins evenly among the three slots yields:

Each portion size is Ⓘ/Ⓘ which implies that $\frac{6}{3} = 2$.

B. $\frac{600}{3}$

We represent 600 by Ⓗ Ⓗ Ⓗ Ⓗ Ⓗ Ⓗ and 3 by

Distributing the six Ⓗ coins evenly among the three slots yields:

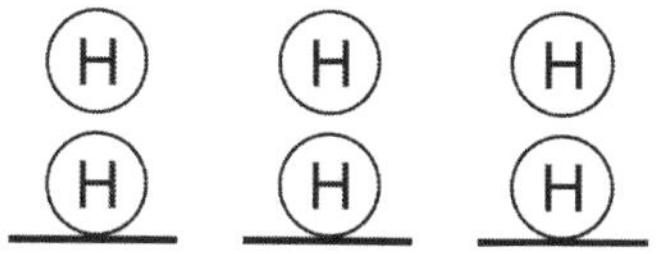

Ⓗ Ⓗ Ⓗ
Ⓗ Ⓗ Ⓗ
___ ___ ___

Each portion size is $\dfrac{Ⓗ}{Ⓗ}$ which implies that $\frac{600}{3} = 200$.

Here, the coins have allowed us to perform the division in grouped form, saving us much effort over representing the collection by 600 unlabeled circles.

Dividing $\frac{936}{3}$ shows how the process plays out when multiple denominations are involved: 936 is represented by:

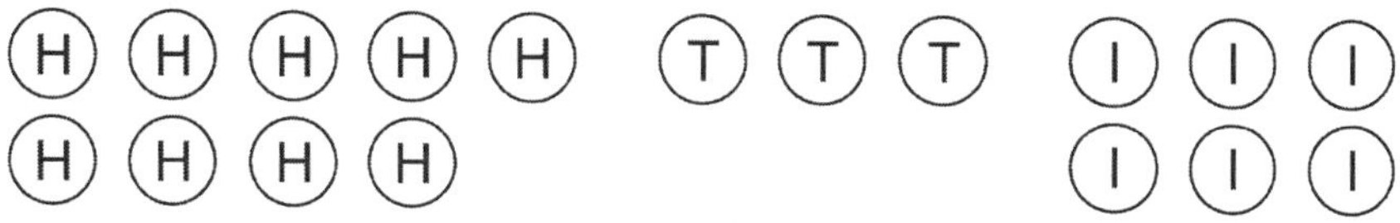

Ⓗ Ⓗ Ⓗ Ⓗ Ⓗ   Ⓣ Ⓣ Ⓣ   Ⓘ Ⓘ Ⓘ
Ⓗ Ⓗ Ⓗ Ⓗ   Ⓘ Ⓘ Ⓘ

Each set of coins is to be equally distributed into the three slots:

_____   _____   _____

We start with the Ⓗs and organize them into groups of three and deal each set to the slots. We have:

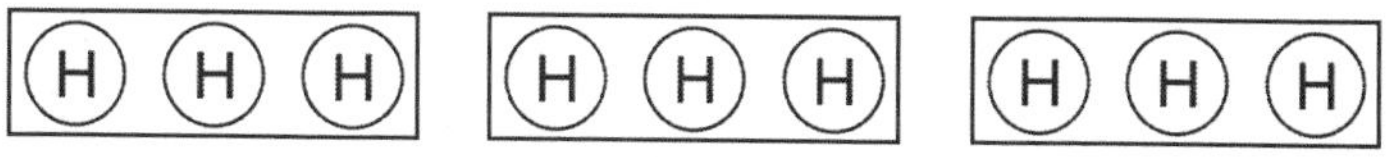

[Ⓗ Ⓗ Ⓗ] [Ⓗ Ⓗ Ⓗ] [Ⓗ Ⓗ Ⓗ]

which deals as:

Ⓗ   Ⓗ   Ⓗ
Ⓗ   Ⓗ   Ⓗ
Ⓗ   Ⓗ   Ⓗ
___   ___   ___

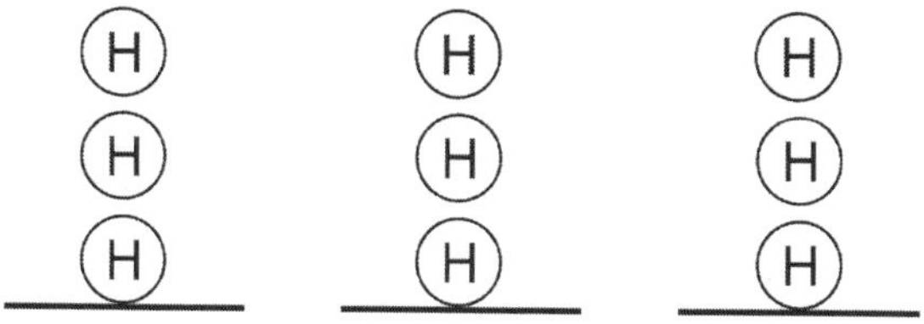

We organize the (T)s into a group of three to obtain 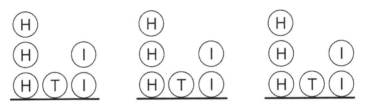. Dealing this set to the pile gives:

H   H   H
H   H   H
(H)(T)   (H)(T)   (H)(T)

Lastly, grouping the (I)s in sets of three gives: . Dealing these to the total gives:

H   H   H
H   I   H   I   H   I
(H)(T)(I)   (H)(T)(I)   (H)(T)(I)

Thus, equally distributing 936 into 3 equal parts gives portion sizes of 312. We conclude then that $\frac{936}{3} = 312$.

In this example, the different coin denominations could have been dealt or apportioned out in any order, with no change in the final result. In other words, since the fit for all three denominations in the three slots was perfect, we could have just as easily dealt out the ones coins first and the hundreds last. However, when the fit for every denomination is not perfect, the most effective way to proceed will be from larger coin denominations to smaller as the example of $\frac{65}{5}$ demonstrates:

In coins, 65 is represented as:

(T) (T) (T) (T) (T) (T)   (I) (I) (I) (I) (I)

These coins are to be equally apportioned into the five slots:

———— ———— ———— ———— ————

We start with the (T)s. Organizing them into groups of five gives:

.

This deals as:

We have the one un-dealt ten and the five ones left over:

. Since there is only one ten coin it can't be distributed over the five slots in its present form. However, if we change denominations and view it as ten one coins its value can be distributed. Doing this and combining it with the five one coins we already have yields:

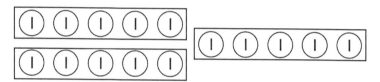

If we organize these fifteen ones into groups of five, we obtain:

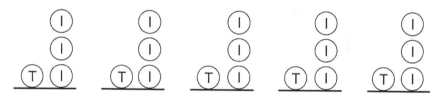

It takes three deals to exhaust the ones:

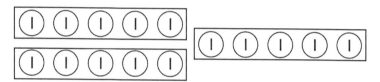

Thus the size of each portion is 13 which means that $\frac{65}{5} = 13$. This can be verified by multiplying 5 times 13. Hopefully, the advantages of proceeding from larger denomination to smaller are clear here (if not, start the process in reverse by first dealing the ones and see if it is more straightforward than the earlier method).

The next example of $\frac{348}{6}$ shows that this method of distributing and changing denominations gives us the answer to a division problem whose answer is not immediately obvious.

348 in coins:

This value is to be equally distributed amongst the six slots:

_____   _____   _____   _____   _____   _____

As before, we start with the highest denomination and work our way down. There are only three (H) coins and these won't fit into six slots so we have to convert the hundreds to tens to obtain:

Organizing the thirty-four tens into groups of six yields:

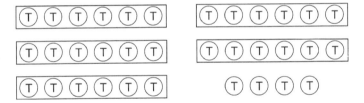

These can be distributed in five deals:

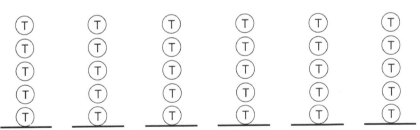

This leaves four tens out of the disbursement. Combining these with the four ones gives:

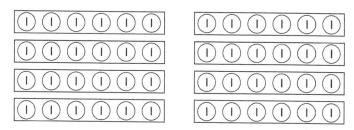

Readying the ones for disbursal by grouping them into packets of six yields:

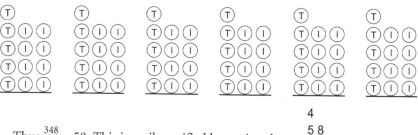

Dealing the eight packets to the slots gives portion sizes of 58:

Thus $\frac{348}{6} = 58$. This is easily verified by noting that

$$
\begin{array}{r}
4 \\
58 \\
\times\ \ 6 \\
\hline
348
\end{array}
$$

To see more examples of division with coin numerals please visit www
.howmathworks.com. This method of division with the coins and slots is
general—meaning that it yields a standard way to perform division in coin
numerals by systematically decomposing the dividend (top number) by its
place values. For any division of a larger whole number by a smaller non-
zero whole number $\left(\frac{\text{larger whole number}}{\text{smaller whole number}}\right)$, we represent the larger number by
place value coins, the smaller number by slots, and use the above methods of
grouping, dealing, and changing denominations where needed, to obtain the
answer (including the remainder, if there is one). It will always work.

In addition to standardizing the process, the method with coin numerals
also offers a clear picture of what is happening conceptually in place value
division. As usual, however, these conceptual methods are too unwieldy for

general use. If we have two or more digits in the divisor (bottom number), we start to encounter tedium—imagine finding $\frac{9234776}{124}$ in the same manner as we did for $\frac{348}{6}$. For this larger division we would ultimately need to distribute more than 3,000 coins of varying denominations among 124 slots. This is simply too much work.

Thus, our next steps include abbreviating coin numeral division. As in multiplication, we will use a mixed system of coins and HA numerals. This mixed system is of a type that mathematical historians call "multiplicative." Such schemes are very much present in the historical record—one of the most notable being the old Chinese numeral system dating back to the second millennium BCE. These systems, as conceptual tools, remain relevant in the twenty-first century.

Ground Game

To abbreviate things, we focus on describing the number of coins of a certain denomination present. Instead of listing all of the coins that represent a number, we will now use a coefficient given in HA numerals. For example, instead of describing nine tens by (T) (T) (T) (T) (T) (T) (T) (T) (T), we will use 9 (T). The "9" is called a numerical coefficient. If we used "nine (T)" instead, then the word "nine" would be our coefficient. If it helps, think of a numerical coefficient as being an adjective that modifies the coin denomination. A couple of examples utilizing numerical coefficients are included in the table:

Pure Coin	Numerical Coefficient and Coin
(T) (T) (T) (T) (T)	5 (T)
(H) (H) (H) (H) (H) (H) (H) (H) (H) (H) (H) (H)	12 (H)

The division algorithm also requires the conversion of higher denomination coins into lower denomination coins. In numerical coefficient language, for example, this unsurprisingly implies the following:

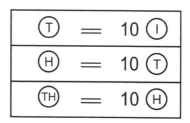

This is simply a restatement of our base-ten grouping in coefficient language. We can also add, subtract, multiply, and divide with these coefficient numerals:

I.   Addition:

$$5 \,\text{(T)} \quad + \quad 8 \,\text{(T)} \quad = \quad 13 \,\text{(T)}$$

II. Subtraction:

$$35 \,\text{(TH)} \quad - \quad 20 \,\text{(TH)} \quad = \quad 15 \,\text{(TH)}$$

III. Multiplication:

$$7 \times 6 \,\text{(I)} \quad = \quad 42 \,\text{(I)}$$

IV. Division:
     Without Remainder:

$$\frac{55 \,\text{(H)}}{11} \quad = \quad 5 \,\text{(H)} \qquad \text{since} \qquad 55 \,\text{(H)} \quad = \quad 11 \times 5 \,\text{(H)}$$

With Remainder:

$$\frac{27 \,\text{(T)}}{7} \quad = \quad 3 \,\text{(T)} \quad R \quad 6 \,\text{(T)}$$

Note that:

$$27 \,\text{(T)} \quad = \quad 7 \times 3 \,\text{(T)} \quad + \quad 6 \,\text{(T)}$$

We also may need to make use of the following:

V. Conversions:

$$85 \ (H) \ = \ 85 \times 10 \ (T) \ = \ 850 \ (T)$$

VI. Combining different denominations:

$$6 \ (TH) \ + \ 35 \ (H) \ = \ 60 \ (H) \ + \ 35 \ (H) \ = \ 95 \ (H)$$

Long Division with Numerical Coefficients

Let's start by revisiting our earlier coin division of $\frac{65}{5}$, this time using a new format:

I.

$$5 \overline{\big)\ 6 \ 5} \quad = \quad \Big|\ \overline{}\ \overline{}\ \overline{}\ \overline{}\ \overline{}\ \ (T)(T)(T)(T)(T)(T) \ (I)(I)(I)(I)(I)$$

The five slots above the bar have replaced the numeral 5.

II. Organize the (T)s into groups of five:

$$5 \overline{\big)\ 6 \ 5} \quad = \quad \Big|\ \boxed{(T)(T)(T)(T)(T)}\ (T) \ (I)(I)(I)(I)(I)$$

III. Dealing the (T)s into the five slots gives:

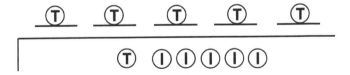

IV. Convert the remaining ⓉⓉ into ①s and arrange into groups of five:

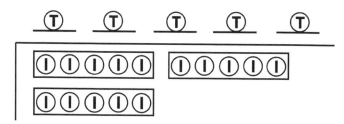

V. Dealing the ①s into the five slots gives:

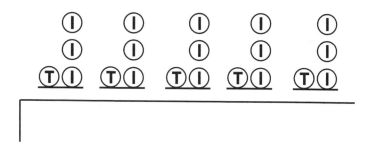

VI. Each portion size is Ⓣ ① ① ①, which gives as before:

$$\frac{65}{5} = 13.$$

Let's now see how the above process translates to numerical coefficient language:

• We first replace 5 ⟌ 6 5   by

$$5\ \overline{\big)\ 6\,Ⓣ \qquad 5\,①}$$

• For II and III, the process becomes divide 6 Ⓣ by 5 (this gives

$$\frac{6\,Ⓣ}{5} = 1\,Ⓣ \quad R \quad 1\,Ⓣ \text{ where,}$$

$$6\,Ⓣ = 5 \times 1\,Ⓣ + 1\,Ⓣ):$$

$$
\begin{array}{r}
1\,\text{Ⓣ} \quad\quad\quad\quad\quad \\
5\,\overline{\big)\ 6\,\text{Ⓣ} \quad\quad 5\,\text{Ⓘ}} \\
-5\,\text{Ⓣ} \quad\quad\quad\quad \\
\hline
1\,\text{Ⓣ} \quad \text{Remainder} \quad\quad
\end{array}
$$

- For IV, the process becomes

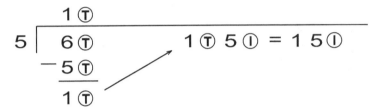

$$
\begin{array}{l}
1\,\text{Ⓣ} \\
5\,\overline{\big)\ 6\,\text{Ⓣ}} \quad\quad\quad\quad 1\,\text{Ⓣ}\ 5\,\text{Ⓘ} = 1\ 5\,\text{Ⓘ} \\
-5\,\text{Ⓣ} \\
\hline
1\,\text{Ⓣ}
\end{array}
$$

- For V, we have that $\dfrac{15\,\text{Ⓘ}}{5} = 3\,\text{Ⓘ}$ and the process translates to:

$$
\begin{array}{ll}
1\,\text{Ⓣ} & \quad\quad\quad\quad 3\,\text{Ⓘ} \\
5\,\overline{\big)\ 6\,\text{Ⓣ}} & 1\,\text{Ⓣ}\ 5\,\text{Ⓘ} = \ 1\ 5\,\text{Ⓘ} \\
-5\,\text{Ⓣ} & \quad\quad\quad\quad -1\ 5\,\text{Ⓘ} \\
\hline
1\,\text{Ⓣ} & \quad\quad\quad\quad \ \ 0\,\text{Ⓘ} \quad \text{Remainder}
\end{array}
$$

- Thus we have that $\frac{65}{5} = 1\,\text{Ⓣ}\ 3\,\text{Ⓘ} = 13$.

Let's next revisit $\frac{348}{6}$ using this new format:

I.

$$
6\,\overline{\big)\ 3\,\text{Ⓗ} \quad 4\,\text{Ⓣ} \quad 8\,\text{Ⓘ}}
$$

II.   Divide the 3 Ⓗs by 6 (this gives 0 Ⓗ R 3 Ⓗ).

$$
\begin{array}{r}
0\,\text{Ⓗ} \quad\quad\quad\quad\quad\quad\quad\quad \\
6\,\overline{\big)\ 3\,\text{Ⓗ} \quad\quad\quad 4\,\text{Ⓣ} \quad\quad\quad 8\,\text{Ⓘ}} \\
-0\,\text{Ⓗ} \quad\quad\quad\quad\quad\quad\quad\quad\quad \\
\hline
3\,\text{Ⓗ} \quad \text{Remainder} \quad\quad\quad\quad\quad\quad\quad
\end{array}
$$

III. Convert the 3 (H)s in the remainder to (T)s :

```
      0 (H)
   ┌──────────────
 6 │  3 (H)        3 (H) 4 (T) =  3 4 (T)              8 (I)
   │ ─0 (H)      ╱
   │  3 (H)
```

IV. Divide the 34 (T)s by 6 (this gives 5 (T) R 4 (T)):

```
      0 (H)                        5 (T)
   ┌────────────────────────────────────────
 6 │  3 (H)        3 (H) 4 (T) =  3 4 (T)              8 (I)
   │ ─0 (H)      ╱               ─3 0 (T)
   │  3 (H)                        4 (T)
```

V. Convert the 4 (T)s in the remainder into (I)s:

```
      0 (H)                        5 (T)
   ┌────────────────────────────────────────────────────
 6 │  3 (H)        3 (H) 4 (T) =  3 4 (T)      4 (T) 8 (I) = 4 8 (I)
   │ ─0 (H)      ╱               ─3 0 (T)    ╱
   │  3 (H)                        4 (T)
```

VI. Divide the 48 (I)s by 6 (this gives 8 (I) R 0 (I)):

```
      0 (H)                        5 (T)                          8 (I)
   ┌─────────────────────────────────────────────────────────────────
 6 │  3 (H)        3 (H) 4 (T) =  3 4 (T)      4 (T) 8 (I) = 4 8 (I)
   │ ─0 (H)      ╱               ─3 0 (T)    ╱              ─ 4 8 (I)
   │  3 (H)                        4 (T)                       0 (I)
```

VII. Thus we have that $\frac{348}{6} = 5$ (T) 8 (I) $= 58$.

Note that since none of the steps in these procedures are marked out, it is possible for all of the steps I–VI to be condensed into one diagram.

We show how this is done with the division $\frac{3924}{9}$:

I.

II. Using a single diagram to perform all of the divisions (progressing from left to right), we obtain:

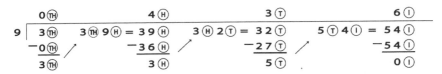

III. Thus we have that $\frac{3924}{9} = 4\,\text{H}\,3\,\text{T}\,6\,\text{I}$ or 436.

IV. As with multiplication on diagrams, alignments abound. In II, if we simply drop the coins on the remainder values when we move them up, we further abbreviate things with no loss of information:

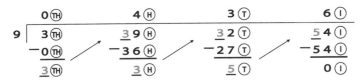

The content of the modern LDA is contained in these coin/coefficient methods. To reach the final stage in this play simply requires that we vertically stack the horizontal steps and jettison the coins.

Long Division with HA Numerals

To show how we get the LDA from the prior situation, we will reorganize the last division $\frac{3924}{9}$. Instead of pushing the remainders up and to the right as we progress through the division, we can just as conveniently push the numerals in the dividend down and to the left like so:

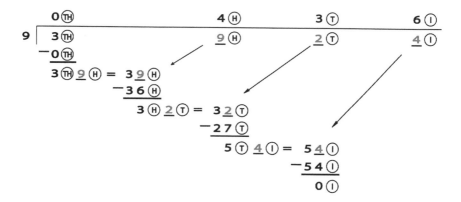

This can be shortened to:

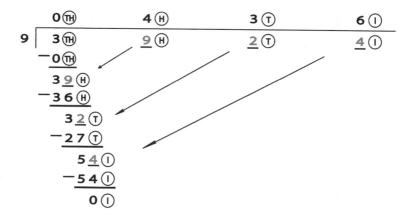

Magic happens now if we jettison the coins and bring the numerals closer together:

```
        0 4 3 6
    9 | 3 9 2 4
       ⁻0
        39
       ⁻36
         32
        ⁻27
          54
         ⁻54
           0
```

This is the modern LDA, where we have explicitly shown the step involving the lead zero. In practice, this is usually left off. In comparing this diagram with the previous one, we clearly see that the change in coin denominations in the hybrid division algorithm translates to vertical steps in the LDA.

We make this explicit in the division of $\frac{6804}{12}$ by labeling the various steps:

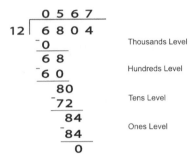

The answer to this division is 567 and it naturally and quite swiftly gives the answer to any of the following questions:

- How many times can we subtract 12 from 6804?
- How many groups containing 12 objects can we form out of 6804 objects?
- What are the portion sizes if we take 6804 objects and distribute them equally into 12 slots?

Each of these questions can of course take on an infinite number of guises including:

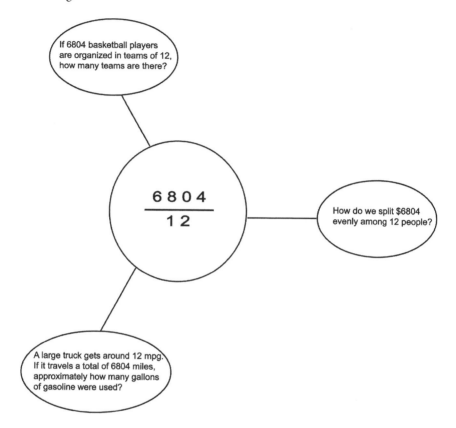

The LDA provides a systematic way of solving division problems. However, as we have already seen, it is not always the quickest way to divide. In many cases, we may do the division faster by canceling out trailing zeros or by simply knowing what number works in one fell swoop (as we did many times in chapter 8).

Even when long division is required it is not always necessary to use every step. For instance, we generally leapfrog the first steps if they yield leading zeros. For example in the prior division, we would generally make a mental note that 12 doesn't go into 6 and immediately go to work on the next step of finding out how many times 12 goes into 68. In this case, our writing of the quotient would not include the lead zero. We may also skip steps by making even bolder guesses (e.g., instead of guessing how many times 12 goes into 68 we might be more daring and try to guess how many times 12 goes into 680).

These are luxuries afforded to us in HA form that are not as available to us when we divide with coin numerals (or even to some extent using numerical coefficients)—demonstrating yet again how HA numerals are truly the Cadillac of all of the numeral systems discussed in this book.

The Great Division

In comparison with our methods for adding, subtracting, and multiplying in HA script, the LDA is a relative newcomer. Prior to 1500, the most common method of dividing with HA numerals was radically different from the LDA. It was known as the galley or scratch method.[3] The following figure shows the method at work in dividing 73,485 by 214:

7 3 4 8 5 |

2 1 4

Before Dividing   After Dividing

The quotient is the number to the right of the bar and the remainder consists of the "unscratched" out numerals on the left side of the bar (reading from left to right). This gives an answer of 343 remainder 83. Don't worry if it looks confusing—it should if you never saw it before. But since the numerals sprout out somewhat evenly around the original diagram as opposed to expanding vertically down from it, the galley method actually takes up less space than the LDA. We won't give the details on how the method works here; to do it justice would take us too far afield. Also, in practice, the numerals were not always scratched out.

The name galley was given to this division because its outline was thought by many to bear a resemblance to the galley ships which dominated the Mediterranean for more than 2,000 years.[4] A particularly ornate example of the division was given in the unpublished *Opus Arithmetica*, a work attributed to a Venetian monk dating back to the sixteenth century:

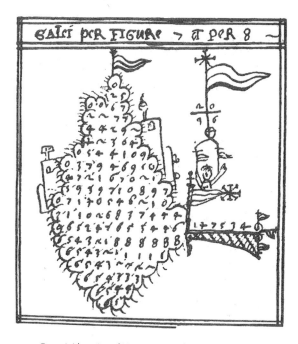

Opus Arithmetica of Honoratus: Mathematical Treasures[5]
Image is from unpublished sixteenth-century manuscript (*Opus Arithmetica of Honoratus: Mathematical Treasures*). Footnote 5 has more information. Image is in the public domain.

If we so wished, we could adorn our own galley division above like so:

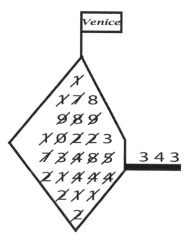

Some teachers in Venice, evidently, made it a requirement that students embellish their galley divisions with such designs.[6]

The galley technique eventually gave way in the 1500s to the method of division called "*a danda*" (the method which gives). Called "the great division" by one Italian writer of the time, it is in essence the LDA we currently use. The word "gives" was used because the procedure involves giving a new numeral to each remainder before starting the next division. In its early form, *a danda* involved placing the quotient on the right as in the galley method but over time the method evolved to a current form in some countries of placing the quotient on the top. This was most likely a result of the arrival in the late 1500s of the decimal notation for fractions which in some renditions of *a danda* required divisions to grow to the right (as more zeros were added after the decimal point) which would lead to interference if the quotient remained on the right.

Most historians in the nineteenth and early twentieth century thought that the galley method originated in ancient India but Lam Lay Yong of Singapore has made a strong case indicating that, while certainly practiced in India in the early centuries CE, use of the technique in China most likely predates this.[7] The origins of our LDA are unknown. It made its first known appearance in print in 1491 in a work by Philipi Calanderi. As controversial among educators as the algorithm is today, it has the peculiar distinction of totally annihilating its chief sixteenth-century rival (although this took some time). For all intents and purposes, the galley method of division is now obsolete. None of the major algorithms for the other three operations that were in place during the sixteenth century can make such a claim.[8] Of the rival methods that exist today for adding, subtracting, and multiplying numbers, many of them were also rivals with each other in the 1500s as well.

Conclusion

Thus concludes our directed study on the conceptual workings of the four elementary operations of arithmetic. Three of them yield to automatic handling once the tables and algorithms have been mastered, with the fourth, division, being effectively tamed with an intelligent guess or two, when needed, to get and keep the LDA rolling.

Division is not the last of the operations of arithmetic, however. Higher order operations, known as exponentiation (or raising to powers) and its inverse known as the taking of roots (e.g., the square or cube root), also exist.

The taking of exponents corresponds to repeated multiplication (when dealing with whole numbers). Thus, if we multiplied $6 \times 6 \times 6 \times 6$, we would abbreviate it as 6^4. The number 6 is called the base (the number being repeatedly multiplied) while the 4 is called the exponent (the number of times we involve the base in the multiplication). If we had 8^{15}, this would mean multiply fifteen eights together.

Can we domesticate repeated multiplication in the same manner as we have the four elementary operations? At the present time, the answer appears to be in the negative, as no simple written grade school algorithms, other than simply performing all of the multiplications, seems achievable. A procedure in writing that allows us to circumvent the repeated multiplications does exist, but it is too sophisticated and messy for the elementary school. Historically, it required the construction of an intricate table that proved exhausting to develop. The Scotsman John Napier, for example, spent nearly two decades of his life, in the late sixteenth and early seventeenth centuries, working on such a table. The logarithms he developed, and that were subsequently improved upon, form the centerpiece in this algorithm. Their use extended and greatly simplified not only the taking of exponents and roots but also the very large and messy multiplications and divisions involving decimals that scientists, such as the astronomer Johannes Kepler, were beginning to encounter.

Surprisingly, logarithms quickly allowed for the creation of a physical device that would far outdistance the abacus in its ability to perform sophisticated computations. Reaching a mature form by the middle of the seventeenth century, the slide rule is a device that can mechanically represent calculations with logarithms in an analogous way that the abacus is a device that mechanically mimics calculations in positional systems. Interestingly, in the case of the abacus, the device preceded the script (HA numerals), while in the case of the slide rule, the script (logarithms) preceded the device.

The question as to who actually invented the slide rule has been somewhat disputed—with the names of Englishmen William Oughtred, Richard Delamain, Edmund Gunter, and Edmund Wingate figuring prominently in the debate. An extremely valuable tool, the slide rule saw its heyday end only relatively recently in the late 1960s and early 1970s with the advent of electronic calculators—but that's a whole other story.

Our approach to arithmetic so far has hopefully demonstrated that mathematics doesn't sit alone in a vacuum, and that its symbols take their forms, in part, based on the needs and limitations of the human beings who developed them. As Raymond Wilder states, "Mathematics was born and nurtured in a cultural environment. Without the perspective which the cultural background affords, a proper appreciation of the context and state of present-day mathematics is hardly possible."[9]

This is true not only of mathematics proper but also of the manner in which mathematics is explained to others. In fact, mathematics education forms its own complex and fascinating subplot in the two vast dramas that comprise its name—mathematics itself and education. Both subjects have rich histories of their own and remain of immense importance today. The same is no less true of their spectacular convergence. It is to this fascinating and important story that we next direct our attention.

III

BEAUTIFUL DREAMS AND HORRIBLE NIGHTMARES

You're an interesting species. An interesting mix. You're capable of such beautiful dreams, and such horrible nightmares.

—Alien sage to Eleanor Arroway in the movie *Contact*[1]

Mathematics education is much more complicated than you expected, even though you expected it to be more complicated than you expected.

—Edward G. Begle, twentieth-century mathematician, math educator, and a chief architect of the New Math[2]

10

Triumph of the Numerals

Not the children of the rich or of the powerful only, but of all alike, boys and girls, both noble and ignoble, rich and poor, in all cities and towns, villages and hamlets, should be sent to school.

—John Amos Comenius (1630s), internationally famous Czech
educator, often considered the father of modern education[1]

The history of education shows us that every subject of instruction has been taught in various ways, and further that the contest of methods has not uniformly ended in survival of the fittest.

—Robert Quick, nineteenth-century British educator,
author of *Essays on Educational Reformers*[2]

RUSSIANS WIN RACE TO LAUNCH EARTH SATELLITE
Saturday Evening, October 5, 1957
Welch Daily News, Mcdowell County, West Virginia

For twenty-two days in October 1957, a shiny aluminum sphere, the size of a beach ball, whizzed through the sky at the breathtaking rate of five miles per second, beeping signals to radio receivers around the globe. Once its transmitter fell silent, the object would remain in orbit for yet another ten weeks before finally plunging to its fiery demise in the upper atmosphere. It was a singular moment in a century packed with them. Sputnik shocked the world! And mathematics education in America would never be the same again.

Talk of serious educational reform had been in the air for more than a decade but few expected this. Now there was the great and real fear, in some cases bordering on hysteria, that America had fallen precipitously behind its Cold War opponent in regards to technical education.

In response, President Eisenhower signed the National Defense Education Act into law in 1958 with the goal of producing a tidal wave of new American-born mathematicians, scientists, and engineers. Just a year later, in Woods Hole, Massachusetts, some of the country's most distinguished scientists and scholars, nearly three dozen in number, met to discuss how best to improve the teaching of math and science in our schools.

Such was the energy and spirit of reform at the time that Yale Professor Ed Begle totally refocused his career from mathematical research to mathematical education by heading up what would become the most well-known of the groups devoted to this reform—the School Mathematics Study Group (SMSG).

Other research mathematicians would accompany him in the quest for reform by joining SMSG or one of the many other consortiums that sprang up throughout the country to address these same issues. *Time* magazine in a September 1961 article stated:

> As secretary of the American Mathematical Society, Begle was in a key spot when Sputnik-stirred mathematicians began to worry about U.S. high schools. They were shocked at "cook book" courses stuffed with unrelated rules, appalled at teachers who themselves hated math. With grants ($4,000,000 so far from the National Science Foundation), Begle organized top mathematicians and teaching experts into five teams, each covering a year of junior or senior high school math. Purpose: to create teachable courses. . . . With the high school books out, Begle plans new texts for kindergarten through sixth grade.[3]

Think of it: research mathematicians, who generally as a community had spurned K–14 teaching, engaged in serious pedagogical discussion about the issues in classroom instruction and math education at all levels.[4] Here were some of the best minds in the subject now hard at work, trying to solve the issues of how best to teach math. For those concerned with such things, what an exciting and energizing time it must have been. It would appear that a golden age in math education had arrived at our shores.

Unfortunately for schools, students, even American society at-large, this golden age failed to materialize. The curricula that ensued from these grand reform efforts became collectively known as the New Math movement, and in most circles today, it is viewed as an effort that, despite such promise at dawn, catastrophically failed to achieve its aims by dusk. If ever there is anything that can serve as a cautionary tale on the multifaceted difficulties inherent in systematically teaching mathematics to large groups of human beings, it is the New Math reform movement of the 1950s and 1960s.

No matter the final tally, however, it is hard to deny that the New Math was clearly a bold stroke to send the math education of minors into a radically new and ambitious direction. The idea might be summarized as an effort to instill in children the unifying notions of abstract mathematical structure: "the goal being to 'make sense' of school mathematics, to anchor an apparently endless series of apparently isolated tricks to a structure within which all these tricks became realizations of a small number of important mathematical principles."[5] A grand unification of sorts. The thinking was that since children learn and take to languages far easier than most adults, perhaps they would do the same with abstract mathematical structure: *Instill these notions in them early and set their natural curiosity on fire!* Now that's a beautiful dream!

A far older but equally beautiful dream in its day was the idea that elementary arithmetic should or even could be systematically taught to all children or even to most adults for that matter. The tale of how elementary arithmetic, one time at the forefront of mathematical thought, became a part of the working vocabulary of the average citizen is a long and involved one easily filling several volumes on its own. In the next two chapters, we will explore some of the highlights focusing specifically on education in the West.

A Question of Speed: The Abacus
versus the Electronic Calculator

We know that abaci were used for centuries to help people overcome the limitations imposed on them by their numeral systems. But how fast could they really be? How would they do arithmetic in comparison with, say, a modern electronic calculator? Believe it or not, in 1946 we found out (see excerpt):

> An exciting contest between the Japanese abacus and the electric calculating machine was held in Tokyo on November 12, 1946, under the sponsorship of the U.S. Army newspaper, the *Stars and Stripes*." In reporting the contest, the *Stars and Stripes* remarked:
>
> The machine age tool took a step backward yesterday at the Ernie Pyle Theater as the abacus, centuries old, dealt defeat to the most up-to-date electric machine now being used by the United States Government. . . . The abacus victory was decisive.
>
> The *Nippon Times* reported the contest as follows:

Civilization, on the threshold of the atomic age, tottered Monday afternoon as the 2,000-year-old abacus beat the electric calculating machine in adding, subtracting, dividing and a problem including all three with multiplication thrown in, according to UP. Only in multiplication alone did the machine triumph.

The American representative of the calculating machine was Pvt. Thomas Nathan Wood of the 20th Finance Disbursing Section of General MacArthur's headquarters; who had been selected in an arithmetic contest as the most expert operator of the electric calculator in Japan. The Japanese representative was Mr. Kiyoshi Matsuzaki, a champion operator of the abacus in the Savings Bureau of the Ministry of Postal Administration.

As may be seen from the results tabulated on the following page [*sic*], the abacus scored a total of 4 points against 1 point for the electric calculator. Such results should convince even the most skeptical that, at least so far as addition and subtraction are concerned, the abacus possesses an indisputable advantage over the calculating machine. Its advantages in the fields of multiplication and division, however, were not so decisively demonstrated. (*The Japanese Abacus: Its Use and Theory*, by Takashi Kojima. Reprinted with kind permission from Tuttle Publishing)[6]

The abacus has since performed well in other arithmetic contests (mostly unofficial) against electronic calculators. Clearly, in the hands of a competent user, the abacus can more than hold its own against modern devices. The film and newsreel company *British Pathe* reportedly has footage of such a contest held in Hong Kong in 1967.

The Counter Abacus

Abaci, like automobiles, however, can and have come in a wide variety of makes and models. The most common view is probably of the type we have discussed: a device with beads to slide up and down on rods (the Japanese Soroban abacus used in the 1946 competition and the Chinese Suan Pan abacus discussed in chapter 3 are both of this type). This hasn't always been the case, however. The abaci that dominated the late medieval European landscape were called counter abaci and they differed in the details from bead abaci.

The counter abacus also has place value built into its design, but instead of beads to represent the amount in a given column, it used tokens or counters.[7] The model shown here is a simpler model than those used in practice and the place value coins have been included for the sake of clarity. Also in contrast to the bead abacus, the counters employed were not attached to the device (see the following).

Table 10.1 Tabulated Results of Abacus Contest (Tokyo, 1946)[a]

Type of Problem	Name	1st Heat	2nd Heat	3rd Heat	Score
Addition: 50 numbers each containing 3 to 6 digits	Matsuzaki	1m. 14.9s (Victor)	1m 16s (Victor)		1
	Wood	2m 0.2s (Defeated)	1m 58s (Defeated)		
Subtraction: 5 problems with minuends and subtrahends of from 6 to 8 digits each	Matsuzaki	1m .4s All correct (Victor)	1m .8s 4 correct (No decision)	1m All correct (Victor)	1
	Wood	1m 30s All correct (Defeated)	1m 35s 4 correct (No decision)	1m 22s 4 correct (Defeated)	
Multiplication: 5 problems each containing 5 to 12 digits in the multiplier and multiplicand	Matsuzaki	1m 44.6s 4 correct (Defeated)	1m 19s All correct (Victor)	2m 14.4s 3 correct (Defeated)	
	Wood	2m 22s 4 correct (Defeated)	1m 20s All correct (Defeated)	1m 53.6s 4 correct (Victor)	1
Division: 5 problems each containing 5 to 12 digits in the divisor and dividend	Matsuzaki	1m 36.6s All correct (Victor)	1m 23s 4 correct (Defeated)	1m 21s All correct (Victor)	1
	Wood	1m 48s All correct (Defeated)	1m 19s All correct (Victor)	1m 25s 4 correct (Defeated)	
Composite problems: 1 problem in addition 30 6-digit numbers; 3 problems in subtraction, each with two 6-digit numbers; 8 problems in multiplication each with two figures containing a total of 5 to 12 digits; 3 problems in division, each with two figures containing a total of 5 to 12 digits	Matsuzaki	1m 21s All correct (Victor)			1
	Wood	1m 26s 4 correct (Defeated)			
Total Score:	Matsuzaki				4
	Wood				1

Results of the contest: Matsuzaki using the abacus, wins 4 to 1 against Wood, using the electric calculator.

[a]*The Japanese Abacus: Its Use and Theory*, by Takashi Kojima.
Reprinted with kind permission from Tuttle Publishing

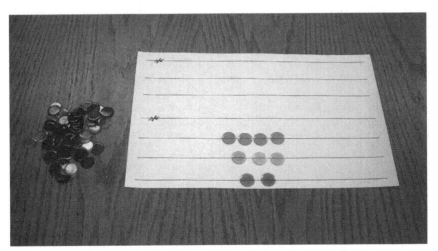

Simple model of a medieval counter abacus with place values included

432 on the counter abacus with extra counters off to the side.
Photo taken by author.

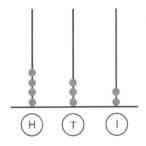

432 on the bead abacus

Woodcut from *Margarita Philosophica* by Gregor Reisch, 1508.
Image is in public domain.

Title page of Jakob Köbel's *Rechenbiechlin* (1514).
Image is in public domain.

Outside of the context of medieval Europe, counter abaci are usually called counting boards. However, since they really are different models of the same underlying idea, and given that writers on the period, including William of Malmesbury (1100s), referred to users of such boards as abacists, we will continue to call the device an abacus.[8] When necessary, we will sometimes also refer to the counter abacus as a line abacus. As with all abaci, to handle addition, subtraction, multiplication, and division on the counter abacus required much skill and memorization.

In practice the counter abacus often had a divider line (with the thousands place asterisk situated on this middle line instead of on the left side as in my simple reconstruction). This allowed two numbers to be represented on the same device for the purposes of adding or subtracting them. The notions of "carrying" and "borrowing" most likely have their origins in this device as these processes were literally played out with the counters both on the board (carrying a counter to the next higher line) and in the pile off of the board (borrowing from the pile).[9] Commerce during the Middle Ages was often done by performing calculations on the counter abacus. The modern day notions of "counter" and "over the counter" in stores and businesses most likely owe their origins also to the counter abacus.[10]

A great disadvantage of an abacus is that it is a device with no memory. Once a calculation is completed, all of the intermediary steps in the calculation are lost. This, as we discussed, led to a situation where the counter abacus was used for calculations between numbers, while Roman numerals were used to record the results of those calculations, especially those relating to commerce.[11] Learning how to use the counter abacus was generally accomplished, much like a novice learns how to cook in a restaurant, through apprenticeship.

Medieval Showdown

On a parallel track, starting in the early centuries CE, came the Indian way of reckoning. It gave people the ability to calculate in writing—meaning in effect that you could "take a photograph" of a calculation. You could even begin to build your own quantitative constitution using objects such as a multiplication table. This "Indian way" in due course would come into direct conflict with the counter abacus, subdue it and its proponents over time, and eventually come to dominate the planet.

This dominance, however, did not happen overnight, not even close, taking nearly a millennium and a half to do so. Why did it take so long? The reasons are complex and varied but they show that mathematics does not stand apart from the limitations and problems of the greater society at-large, nor from human fallibility. This is true even today.

It is worth noting that for all of their marvelous accomplishments in mathematics and science, the ancient Greeks totally missed out on discovering this type of numeration system. The great eighteenth- and early nineteenth-century French mathematician Pierre Simon Laplace states:

> It is India that gave us the ingenious method of expressing all numbers by means of ten symbols, each symbol receiving a value of position as well as an absolute value; a profound and important idea which appears so simple to us now that we ignore its true merit. But its very simplicity and the great ease which it has lent to computations put our arithmetic in the first rank of useful inventions; and we shall appreciate the grandeur of the achievement the more when we remember that it escaped the genius of Archimedes and Apollonius, two of the greatest men produced by antiquity.[12]

Mentioning the fact that something so basic and fundamentally important to all of mathematics could escape the notice of the great Greek geniuses is not to criticize them. It was done to illustrate that when people don't perceive that there is an issue to even worry about, they generally won't devise a new way to "fix it"—no matter how smart they are. The majority of the problems that the ancient Greeks were concerned with simply did not, for them, involve devising a better system of numeration.

Could possession of HA numerals have helped them in their investigations? Without question, but for where the interests of their intellectual traditions and culture lay, the tools they had must have felt adequate enough for them.[13] Put another way, the Greeks weren't sitting in their homes pining away for a better number system any more than they were bemoaning the fact that they didn't have automobiles. Not having the luxury of our modern viewpoint, they simply did not see this situation as a major issue. The same can undoubtedly be said in defense of other ancient peoples as well—particularly, in regards to their not employing formal proof (as we view it today) or developing a fully symbolic algebra, and so on.

These discoveries in numeration radiating out from India passed on into the medieval Arab world to successfully compete with two other rival systems of reckoning (finger reckoning and sexasgesimal numerals).[14] The Arabs were eventually to make important and everlasting discoveries in many areas of mathematics, particularly algebra (even giving us the very word itself). Just how outstanding and farsighted many of their contributions were is increasingly coming to light and it was through contact with the Arabs that medieval European mathematicians slowly emerged from the mathematical stagnation that existed on the continent during this period.

Two important European players involved in the transmission of the In-dian way of reckoning were Gerbert d'Aurillac (ca. 946–1003) and Leonardo of Pisa (Fibonacci) (ca. 1170–1250). Since the numerals were used in the Arab world, they were at first called Arabic numerals by some Europeans. Others such as Fibonacci, however, knew that these numerals had their origins in India and over time they eventually came to be called the HA numerals, which is how we have referred to them in this book.

Gerbert, who become Pope Sylvester II, used the numerals in a hybrid fashion. He used a counter abacus but, instead of using unmarked tokens, he fashioned special types of tokens by marking them with HA numerals and representing the place values with Roman numerals. Thus, using modern nu-merals, Gerbert would have represented 432 on his counter abacus as:

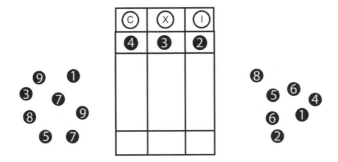

Contrast this with the way 432 was represented with unmarked tokens on the line abacus. These coins, called apices, were also not attached to the board and could be replaced by others as needed. Our process of crossing out num-bers in a subtraction could have, in Gerbert's system, amounted to one apice being removed from the board and replaced by another.

Despite its earlier appearance, Gerbert's abacus was eventually replaced by the line abacus. However, Gerbert's use of the HA numerals on the apices did serve to give these symbols a greater familiarity in the regions that they were used, but unfortunately, his introduction was not the definitive introduction of these numerals into Europe. That would come more than two centuries later.

Leonardo of Pisa

By the High Middle Ages, a revolution in commerce had taken the towns of northern Italy by storm. This revolution would create an influential new class of citizen whose clout came not from religious ties or noble birth but from money. So enabling was this liquid wealth for towns such as Venice, Pisa, and Genoa, that they became independent and powerful city-states—able to, in alliance, meet the Holy Roman Emperor Frederick Barbarossa on his own terms in 1176 CE and force him to the peace table.[15]

As commerce grew to become more regional and international, the complexity grew in terms of the types of problems that merchants were being called upon to handle. An example of the types of problems merchants were concerned with in 1202 is given here:[16]

- *Finding the value of merchandise based on another sale*: If one hundred rolls are sold for forty pounds, what can I buy for two-and-a-half pounds?
- *Currency exchange in a complex economy*: If one imperial soldo is worth thirty-one Pisan denari, how many imperial denari can be exchanged for eleven Pisan denari (one imperial soldi is equivalent to twelve imperial denari)?
- *Partnerships*: Two men form a company in which the first man puts in eighteen pounds of some currency and the second puts in twenty-five pounds. If the company makes a profit of sixty-five pounds, how much of the profit does each man hold?
- *Mixing alloys*: A certain man has nine pounds of silver. If he wants to make money consisting of two ounces of silver in each pound, how much copper should he add to the silver to make this happen? Note: twelve ounces here make a pound.

Amidst this backdrop lived Leonardo of Pisa, often hailed as the most influential European mathematician of the Middle Ages. He learned of the HA numerals during his boyhood tenure with his father in Algeria and was so taken with them that in 1202 he decided to write what would turn out to be one of the most important and influential math books of all time, *Liber Abaci* (Book of Calculation), which showed how to use these numerals, in written, not "token/apice" form. The title of this book is misleading as it is a book about calculation in writing not calculation on the abacus.

His intent was to convey the absolute magic possessed by this numeral system out of Asia. In a real sense, he was one of the most important heralds of the fact that elementary arithmetic could be fully and conveniently accomplished in script alone. Merchants in the powerful city-states were impressed, and in due time the two-pronged process of calculating on the abacus and then recording the results in the accounting books with Roman numerals began to give way to the process of both calculating and recording in pen using these new numerals.[17] Books, such as *Liber Abaci*, which discussed how to calculate with HA numerals in writing, became known as "algorisms."

A few passages from chapters 1 and 11 of the thirteen-century *Liber Abaci* (translated into English by Laurence Sigler 800 years later in 2002) are provided here:

Chapter 1
Here Begins the First Chapter

The nine Indian figures are:

$$9 \quad 8 \quad 7 \quad 6 \quad 5 \quad 4 \quad 3 \quad 2 \quad 1$$

With these nine figures, and with the sign 0 which the Arabs call zephir, any number whatsoever is written, as is demonstrated below. A number is a sum of units, or a collection of units, and through the addition of them the numbers increase by steps without end. First, one composes from units those numbers which are from one to ten. Second, from the tens are made those numbers which are from ten up to one hundred. Third, from the hundreds are made those numbers which are from one hundred up to one thousand. Fourth, from the thousands are made those numbers from one thousand up to ten thousand, and thus by an unending sequence of steps, any number whatsoever is constructed by the joining of the preceding numbers.

Chapter 11
Here Begins Chapter Eleven on the Alloying of Monies

When it is made up from mixed silver and cooper, no matter what is the face value, it is indeed called money. However, money is called major when a pound of it contains more silver than copper, and it is the more desired. Minor money truly is when there is less silver. It is called alloying money when some given quantity of silver is put in a pound of money. And when we say, I have money with any number of ounces, as when we say with 2, we understand that in a pound of the money are had 2 ounces of silver. Money is alloyed indeed in three ways. The first way is when it is alloyed from a given quantity of silver or copper. The second is when it is alloyed from any given monies with the addition of silver, or copper, or both. The third is when it is alloyed only from given monies. And all are contained in this complete chapter which we separate into seven distinctions.

Liber Abaci, by Leonardo of Pisa (1202), translated from the Latin by Laurence Sigler in Fibonacci's *Liber Abaci,* 2002, 17, 227.
Reprinted with kind permission of Springer Science+Business Media.

Liber Abaci was written in Latin but by the end of the thirteen century, abbreviated Italian versions of the book began to appear.[18] The influences of this, and other texts written along similar lines, were evidently so far-reaching that their methods and spirit began to play a prominent role in an entirely different kind of school—the merchant vernacular school (reckoning or calculation school).

After completing elementary school, Italian children of primarily well-situated families now had a choice of attending either a Latin grammar school or a merchant vernacular school. In contradistinction to the grammar schools, which focused on teaching Latin to prepare for a career in medicine, law, religion, or perhaps the university, vernacular schools taught reading and writing in Italian as well as accounting and commercial mathematics using HA numerals to prepare for a career in business.[19]

The technique of using the new numerals to solve commercial problems came to be called *abbaco* (often called "abacus" by later writers) and sometimes the vernacular schools themselves were called *abbaco* schools. Famous alumni of *abbaco* schooling likely include perspective painter Piero della Francesca, mathematicians Luca Pacioli and Niccolò Fontana (Tartaglia), polymath Leonardo da Vinci, and the philosopher Niccolò Machiavelli.[20] The echo of these schools on the course of Western mathematical education can still be heard to this very day.

While students had to first be taught the basics of the HA numeral system and how to perform the fundamental operations of arithmetic, the primary goal of the abbaco schools was to teach students how to use this newly acquired facility in numeration to solve the commercial problems of the day. Most instruction in abbaco seemed to start sometime after the age of ten and lasted for about two years.[21] After this the student would enter a long apprenticeship on the job, gradually working their way up through the business.

Evidently, the instructional method involved the teacher discussing a particular problem along with its solution while the students faithfully copied it all down in their copy books. Over the course of their studies, the students would amass a large number of specific problems with procedures for their solutions (performing everything, including the calculations in writing allowed for this). Years later when they encountered a specific problem on the job which they could not immediately solve, their copy book would serve as a valuable reference, as they could scan through it for the same type of problem and once found employ the same method of solution.[22]

In spite of the great strides made in Italy using the HA numerals, use of the counter abacus did not immediately disappear in the rest of Europe or even in Italy for that matter. Which meant that, by the late thirteenth century in Europe, two rival methods of reckoning existed: one performing calculations using the counter abaci then recording the results in Roman numerals and the other performing calculations and recording the results both in writing using the HA numerals.

Practitioners of the first method were called abacists while practitioners of the latter were called algorists. These systems would spread (in the case of HA numerals) and disappear (in the case of counter abaci) at differential rates throughout Europe, existing side by side, in some cases, right on up until the arrival of the printing press in the fifteenth century and well into the sixteenth.[23]

Knowledge Explosion: The Printing Press

Now we go back to the question: Why did it take so long for the HA numerals to be universally accepted across Europe? The reasons most likely include aspects of each of the following:

- The production of books in sufficient quantity explaining the numerals was difficult until the advent of the printing press.
- The shapes of the numerals were not completely standardized.
- Even in the cases where the numerals were standardized, mistakes in recording them could still happen leading to potentially catastrophic interpretations (see chapter 3, note 8).
- The Black Death pandemic of the fourteenth century which caused the death of perhaps more than half of the population of Europe.
- Materials to be written upon, such as paper, a Chinese invention, were expensive and not readily available.
- Basic resistance to change including opposition from abacists and conservative merchants under the old system.
- The numerals were often associated with mysticism and astrology.
- The conceptual difficulties in understanding the meaning of the number zero.[24]
- Learning how to use the system took a lot of effort. This is not a trivial thing for we still deal with similar circumstances today in America. The metric system is clearly a superior system to the one we use (for instance, it allows conversions between say units of length or weight [mass] to be as easy as conversions with money; that is, converting say 532 cents to $5.32 by simply moving the decimal point two places) yet due to its unfamiliarity, we refuse to change to it.
- The numerals were looked down upon initially by universities as commercial/merchant arithmetic and not a suitable topic for their attention. The universities focused more on arithmetic as a theoretical science

rather than a practical art studying such things as prime numbers, square numbers, and perfect numbers in preference to studying algorithms for multiplication and division. On those occasions where they did deign to use the HA numerals, it was mainly for calculating the dates of Easter and other religious festivals.[25]

- Books written about these numerals were often written in Latin as opposed to the vernacular or native language and this limited access.
- It would be many years, after the founding of abbaco schools, before mathematics found equal prominence in the curricula of educational programs throughout the rest of the continent.

Thus the odds were actually well stacked against the quick universal acceptance of these numerals in Europe. But accepted they did eventually become, as subsequent outbreaks of plague never approached mid-fourteenth century levels, respect for the power of numerical reasoning to successfully aid in business ventures continued to grow, paper, while still not cheap, became more available, resistance from the universities began to cool, and perhaps most importantly of all the mechanical printing press was invented around 1440 by Johannes Gutenberg. The significance of this most important of inventions is hard to overstate:

> What gunpowder did for war the printing press has done for the mind.—Wendell Phillips[26]

> The printing press . . . soon did for knowledge what steam has done for trade: reduced time and distance to their lowest terms in the intellectual commerce of the people. Men no longer had to make long and weary pilgrimages to the homes of learning: knowledge was brought to their very doors. Often with less trouble than was taken, formerly, to teach one pupil by the voice, a teacher now taught thousands by the pen.—Henry Holman[27]

This revolution in the spread of knowledge created by the printing press can be likened to the revolution in the spread of energy use affected by the production of electricity at generating stations. Before the late nineteenth century, most energy use had to be close to the sources where the energy was produced. With the development of power plants, however, this would all change, allowing for the creation of vast electrical networks that gave the capability of transporting energy hundreds of miles from the places of production with spectacular results—lighting up and powering entire cities from afar.

The printing press turned some thinkers into true best-selling authors and celebrities. Dutchman Desiderius Erasmus (1466–1536) reportedly sold more than 750,000 copies of his work during his lifetime, while more than 300,000 copies of Martin Luther's work sold in just three short years, from 1518 to 1521—providing ample fuel for the fires of reformation.[28] By 1600, an estimated 150 to 200 million volumes had been collectively produced by all the printing presses in Europe.[29]

Printed books on arithmetic also shared in this revolution, albeit at a slower pace. By the end of the 1400s it has been estimated that around thirty different texts on elementary arithmetic had been printed, and by 1750 at least three thousand more.[30] The first known such text is the *Treviso Arithmetic* printed in Treviso, Italy, in 1478 (the first translated passages are included here).[31] It is worth mentioning that this book was written in the local Venetian dialect, and not Latin, which opened it up to a much wider audience.

At Treviso, on the 10th day of December, 1478.

Here beginneth a Practica, very helpful to all who have to do with that commercial art commonly known as the abacus.

I have often been asked by certain youths in whom I have much interest, and who look forward to mercantile pursuits, to put into writing the fundamental principles of arithmetic, commonly called the abacus. Therefore, being impelled by my affection for them, and by the value of the subject, I have to the best of my small ability undertaken to satisfy them in some slight degree, to the end that their laudable desires may bear useful fruit. Therefore in the name of God I take for my subject this work in algorism, and proceed as follows:

Source: First passages of the *Treviso Arithmetic* (1478) translated into English by David Eugene Smith, 1911.

Seeds of Universal Education

By the 1500s in Europe, there was in place a mechanism for spreading knowledge to thousands of people. Renaissance humanists sought to educate a new class of leaders and professionals according to the lofty heights of ancient Greece and Rome. With the Protestant Reformation came the desire to teach large masses of common people how to read in their own languages so that they could read the Bible for themselves.

At this time we witness the rise of a great wave of educators and educational thinkers whose ranks include: Juan Luis Vives, Francois Rabelais, Johannes Sturmius, Petrus Ramus, Michel de Montaigne, the Jesuits, Richard Mulcaster, Wolfgang Ratke, and John Amos Comenius. These individuals and groups as well as many more began to grapple with some of the eternal issues anyone must wrestle with when considering educating large and diverse groups of people. They include:

- what material to teach
- who to teach it to
- the purpose in teaching it
- how to teach it
- who will do the teaching
- how to assess the results of teaching it
- how to effectively administer the whole enterprise

The initial social and economic circumstances, which led to an increase in the number of people who received schooling as well as slowly changing views on whether the masses should also be educated, would help elevate the issues involving education into one of the outstanding global concerns for the rest of the second millennium CE and beyond. By the early twenty-first century, at the center of this education storm in many countries would sit science and mathematics.

The sixteenth century saw the gradual defeat of the counter abacus by the HA numerals. This defeat was first felt in Italy and Spain, actually occurring earlier in these countries, and gradually creeping northward throughout the 1500s. The algorisms printed after Gutenberg's invention serve as a trail to the thinking on elementary arithmetic at the time. Most of these books had huge sections dedicated to solving the commercial problems of the day. This is not surprising since, as we have seen, the commercial influences on arithmetic education were enormous, predating even Fibonacci's time.

Unfortunately, by the seventeenth and eighteenth centuries this influence would take a decided turn for the worse as the number of people seen needing arithmetic swelled to include the ranks of those whose backgrounds and interests were very far and away from those of the sons of rich Italian merchants of the thirteenth through fifteenth centuries immersed in a robust commercial tradition. This would be true throughout all of Europe and ultimately by transfer to the emerging republic in America.

What Is So Special about the HA Numeral System?

After emerging the victor in Europe following a struggle of more than four hundred years, the conquest of HA numerals over other systems of reckoning would continue unabated, eventually engulfing the entire globe by the twenty-first century. The many nations of the world still practice different religions, still use different monetary systems, still speak with different sounds, still write using different letters and glyphs, and still live under different brands of government, but when it comes to numeration they all calculate in some measure using the same fundamental set of numerals.

A clean sweep! Why did it happen? What do the HA numerals really offer us that make them so special?

They unquestionably offer a massive speed advantage in calculation versus Roman numerals and our coin numerals viewed purely as written systems. But is speed the only reason why they dominate today?

They certainly don't offer a speed advantage over the bead abacus—which we know can be as fast as modern calculators. And while performing arithmetic in writing with HA numerals may actually have rivaled the counter abacus in terms of speed, that alone wouldn't have been great enough to overthrow the device.[32] The truth be told, as far as bare calculation and registering the results alone went, the two-pronged system, of using Roman numerals to record and the counter abacus to compute, was workable. To effect such a massive change across the globe something else must have been at play. If not, it is doubtful whether users of the unquestionably swift bead abaci, such as the Chinese and the Japanese, would have ever adopted HA numerals.

It is important to remember, the HA numerals didn't simply replace a set of numerals or a physical device. They forever altered clever systems that employed the two in powerful combination (totally overthrowing the one practiced throughout medieval Europe). There must be something extra special about combining both the representation and calculation of quantity into a single written system that more than makes up for the losses endured in abandoning centuries old and familiar systems

The answer to exactly what that something special is certainly deserves the attention of a cognitive scientist or two. Surely it must touch upon some of the general advantages that written communications can offer over kinetic communications (e.g., spoken language or calculations on an abacus). By being a full-service set of numerals, the HA script allowed for permanence and static visibility of expression in a small space, making possible the literal explosion of a "literature of quantity."

In discussing the Italian merchant schools, we mentioned that their copy books allowed students to create large reference books complete with solved problems that they could use repeatedly to deal with the problems they later encountered on the job. This ability to record and catalog hundreds of problems complete with their solutions in a single reference book is seamless with compact written HA numerals.

How would you efficiently accomplish this in the two-pronged system where the hard calculations were done on the abacus (with each step on the device being literally obliterated by the next one in a calculation) and only the results were written in Roman numerals? Robert Recorde in his work, *The Grounde of Artes* (1543), took seven pages of text just to show how to multiply 1542 times 365 using the counter abacus.[33] It would be difficult for many to follow dozens of such examples a few days later, much less many years later.

On the other hand, by spectacularly merging the calculation and representation of quantity together in a single system of writing, the HA numerals allowed for the visible preservation of both. Moreover, the numerals were able to miniaturize the entire business of numeration to such a degree that the whole process could often be taken in with a single glance. The removal of the abacus as an intermediary also meant that you could more easily do mental calculations (as a quick check) directly from the printed page (swiftly obtaining a ballpark estimate of what 51×72 is by substituting the much easier 50×70 in its place). This was not an insignificant capability to those regularly handling business transactions.

No two-pronged system could come close to approaching this sweeping clarity of view complete with the ability to easily capitalize on all sorts of recognizable patterns (e.g., the ability to tell that a collection containing 543,212,896,758,365,967,806 objects can be split into two equal parts [it's an even number], not by first experimenting with the actual objects, but by simply knowing that the HA string describing it ends with a 6).

Whereas other numeration systems allowed for the creation of a specialized and limited craft to do arithmetic (involving calculations by the specialist on the abacus or some other device), the HA system, by capturing the calculations and motions of arithmetic on paper, allowed for the creation of an industry. This industry would ultimately remove much of the mystery and aura surrounding elementary calculation, eventually paving the way to making arithmetic accessible, in principle, to everyone (think of the modern-day revolution affected by shrinking computers from room-size machines available, at great expense, only to business and governments, to the iPads, desktops, and notebooks of today—placing the wonders of electronic-powered computation at the fingertips of millions).

Moreover, this system was capable of natural extensions to other classes of important numbers such as the integers, fractions, and irrational numbers. Finally, the numerals laid the groundwork for the complete development of the mathematical marvel of the late 1500s and early 1600s: the fully written symbolic algebra—the importance of which is hard to overstate.

And so the HA numerals march on. From the early centuries CE to the Middle Ages to the Renaissance to the present day, they still impress. Without a doubt, they continue to be most deserving of the praise heaped upon them.

Are they without peer? Out of all of the cultures of the past, were the ancient Indians the only ones to devise such a system? We now know of at least three other cultures that constructed positional systems of some kind. The origins of all three may predate the origins of the HA system. Credible suggestions have even been made in the case of one, that the ancient Indians might have obtained crucial ideas for their numerals from that system.[34] Whether this is true or not remains to be seen, but if so it only magnifies the wealth coming out of Asia—it does not diminish the achievement on the Indian subcontinent of fusing together representation and calculation into one grand single written system of numerals. We now take a brief look at each of these systems.

Other Positional Systems

In this section, we will simply give data on each system along with an example:

1. Sumerian/Babylonian Cuneiform Numerals:

Type:	A sexagesimal (base-60) positional system
	59 distinct symbols (without a zero)
	Groupings occur in packets of sixty

Dates of Major Usage: Third millennium BCE to the Renaissance (1500s)

Symbols:

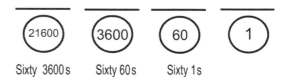

Place Values:	Sixty 3600s	Sixty 60s	Sixty 1s

Other features: (1) Numerals were written horizontally; (2) This system is the earliest known positional numeral system; (3) System did not possess a real zero, meaning that the size of the number represented often had to be inferred from the context; (4) Used by astronomers up until the Renaissance; (5) Remnants of this system survive to this day in time keeping (60 minutes, 60 seconds) and angle measurement.

82 in Cuneiform: With the coin tags:

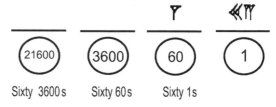

Without the coin tags: Υ ⟪𝍝

Translation: $1 \times \boxed{60} + 22 \times \boxed{1} = 60 + 22 = 82$

Helpful hint: Think of this as 82 minutes being equivalent to 1 hour and 22 minutes.

2. Mayan Numerals:

Type: A vigesimal (base-20) positional system
20 distinct symbols (including a zero)
Groupings occur in packets of twenty

Dates of Major Usage: Early centuries CE to Spanish Conquest (1500s)

Symbols:

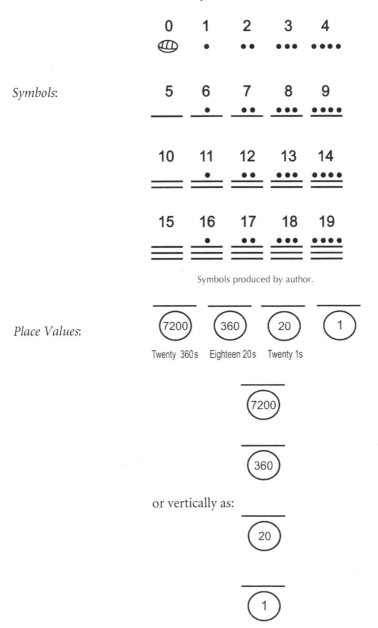

Symbols produced by author.

Place Values:

or vertically as:

Other features:

(1) Numerals were written vertically. (2) System was not pure base 20 as the third place value should be 400 (twenty 20s) instead of 360 (eighteen 20s). It is believed that this position was

chosen to have the value 360 due to the fact that the system was designed mainly for calendar computations. (3) System had a real zero.

82 in Mayan:

With the coin tags:

Without the coin tags:

●●●●
●●

Translation: $4 \times \textcircled{20} + 2 \times \textcircled{1} = 80 + 2 = 82$

Helpful hint: Think of forming 82 dollars from four twenties and two one dollar bills.

3. *Chinese Rod Numerals* (different from the old Chinese numeral system mentioned in chapter 9):

Type: A decimal (base-10) positional system
9 distinct symbols (without a zero)
Groupings occur in packets of tens

Dates of Major Usage: Fourth century BC (Warring States) to the Ming Dynasty (fourteenth–seventeenth centuries)

Symbols: There were two sets of symbols for 1, 2, 3, . . . , 9. For the sake of readability, these were used in alternate positions, the first set for units, hundreds, ten thousands, etc., and the second set for tens, thousands, hundred thousands, etc.

Units, Hundreds, Ten Thousands,…

Tens, Thousands, Hundred Thousands,…

Place Values:

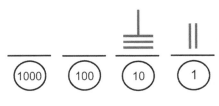

Other features: Numerals were not written but were tokens or chips much like Gerbert's apices and were displayed horizontally on a counting board. Early algorithms in this system closely resembled some early algorithms in Arabic texts using HA numerals.[35]

82 in Rod Numerals: *With the coin tags:*

Without the coin tags:

Translation: $8 \times \circled{10} + 2 \times \circled{1} = 80 + 2 = 82$

Conclusion

Our story is far from finished. The tale of arithmetic education in the Western world doesn't end simply with the triumph of the HA system in Europe during the late Renaissance. Actually, this is precisely where it gets really interesting. It brings to mind the words of Ed Harris, in the movie *Apollo 13*, when in character as NASA flight director Gene Kranz he states, "Gentlemen, we've given our guys enough to survive til reentry. Well done. Now we gotta get 'em in."[36]

Now that many of the essential pieces were in place, the problem became, how do you actually do it: get the know-how contained in elementary arithmetic into the working vocabulary of the millions we seek to educate? Hundreds of years later, in spite of truly spectacular successes, we have yet to reach a consensus on the answer to this fundamental question.

11

From the Frontier to the Classroom

The wits and endeavors of the world are so many scattered coals or fire-brands, which for want of union are soon quenched, whereas being but laid together they would yield a comfortable light and heat. . . . So we see many wits and ingenuities lying scattered up and down the world, whereof some are now labouring to do what is already done, and puzzling themselves to re-invent what is already invented. Others we see quite stuck fast in difficulties for want of a few directions which some other man (might he be met withal) both could and would most easily give him.

—Sir William Petty (paraphrased), seventeenth-century
English economist, scientist, and philosopher[1]

Human beings, who are almost unique in having the ability to learn from the experience of others, are also remarkable for their apparent disinclination to do so.

—Douglas Adams, twentieth-century author
of *The Hitchhiker's Guide to the Galaxy*[2]

HUMAN BEINGS HAVE BEEN busy for the last 10,000 years! Our task, a noble one at that, is just this: to take a wide swath of an ever expanding mountain of knowledge, compress it into small, digestible packages, and present it to hordes of students over the short span of an educational curriculum—all the while hoping that their minds will take to it and that society will be the better for it.

Are you kidding me? With so much to consider, so many different avenues to take, so many people affected, so much bureaucratic red tape, and so much attitude to boot—and with stakes that simply could not be any higher, is it fair to set this charge to anyone? How to successfully negotiate such treacherous waters has always been one of the major dilemmas for formal education.

During the sixteenth, seventeenth, and eighteenth centuries, as the variety and number of schools continued to grow, European educators were faced with the complex task of incorporating thousands of more young people into the educational scheme of things.[3] How best to do this? How should we teach children? How do they really learn things?

Should they be educated as a small, less sophisticated, and imperfect type of adult? Or should we treat them as a different class of learner entirely—a class which has its own way of seeing the world?

More times than not, it was the former view that was accepted. This was certainly the case in arithmetic. The early printed books on HA arithmetic were written for adults or advanced youths in a one-size-fits-all fashion, and were based, in part, on the commercial tradition of the Italians. The early printed school textbooks that followed continued in this vein—leading to the problematic situation that textbooks, born of the commercial spirit of the thirteenth–fifteenth centuries and written in a style more suitable for adults and numerically savvy children from merchant families, were still being used to educate large numbers of new children whose backgrounds would vastly differ from their predecessors', and whose interests and needs were either new to commerce or would lie outside of it.

During these years, it was also common for instructors to apply corporal punishment in liberal doses, not only to the misbehaved mind you, but also to children who were not learning the material at the "appropriate pace." Multiply 543 × 25 correctly, or else![4]

Thus already in its infancy, the growing idea of universal education was faced with the makings of a nightmare in regards to mathematics instruction. And perhaps, with the exception of the overly harsh discipline, it is hard to cast blame—the problems of mathematics education are just plain complicated and tricky, leading those who don't appreciate this fact and who may only have the noblest of intentions nevertheless to create true chimeras.

The HA numeral system, as well as mathematics in general, has always presented educators with a dilemma. Its ability to compress so much work into a compact written form is one of the reasons why we continue to sing its praises and why it won the day over all the others; yet, as we have seen, this very ability to compress information in writing turns out to also be the ability to mask what is truly going on conceptually.

Unfortunately, the majority of those involved in European education at the time did not recognize this as a serious enough concern, and ushered into place systems that would ultimately lead a large percentage of individuals who were taught mathematics in a classroom setting to not understand it at all or at the very least to develop an intense frustration or dislike of it.

In this chapter, we zoom in on England and America. The methods of education that ensued in England during this period were the ones that were eventually adopted in colonial America and their effects can still be felt today in American attitudes toward mathematics education.

Arithmetic Education in England: Seventeenth and Eighteenth Centuries

The exciting changes occurring in the development and spread of knowledge in arithmetic and algebra, spearheaded in southern Europe, slowly crept their way northward. By the 1520s, this resulted in the printing of Englishman Cuthbert Tunstall's (Tonstall's) book, *De arte supputandi*. Unfortunately, the text was written in Latin and appears not to have had much influence on later English educational thought, despite the high respect enjoyed by Tunstall during his lifetime.

Within twenty-five years, more books followed (this time written in English) including: the anonymous *An Introduction for to Lerne to Rekyn with the Pen and with Counters* (1537) and *The Grounde of Artes* (1543) by Robert Recorde (the man responsible in another work for introducing our present-day symbol for equality " = "). Both were to have great influence and success; each being printed in many editions over the next hundred years. It is Recorde's work, however, which has received the lion's share of the praise historically, but evidently the 1537 work also exerted an influence, as Isaac Newton (1642–1727) himself was found to have a copy of *An Introduction* in his collection but not *The Grounde*.[5]

One of the major goals of these books, and others that followed in the latter 1500s, was to introduce a completely new system of numeration to people more familiar with Roman numerals; and their audience was anyone who cared to learn about it. This in effect meant that the works served the purposes of both a textbook and a popularization.

As knowledge of HA numerals gradually trickled down into the classroom, one would suspect that textbooks specifically written to address an audience of young people would soon follow with explanation and demonstration being made to play a prominent role. Shockingly, however, nothing like this would happen in England for hundreds of years!

By the mid-1600s, the basic textbooks on arithmetic being used were almost completely devoid of any real explanation of the rules (see Appendix A). They were terrifyingly brief. The *Treviso Arithmetic* (1478) itself, concise as it was and written for those already familiar with arithmetic, gives far better explanations of the rules than many of these seventeenth- and eighteenth-century textbooks. How could something like this happen?

According to the celebrated nineteenth-century English mathematician, Augustus de Morgan, it was due "to the commercial school of arithmeticians," who came to the fore after Recorde, that were to blame for the "destruction of demonstrative arithmetic" in England or at least "the prevention of its growth."[6] According to math historian Florian Cajori, the blame was due to the authoring of arithmetic textbooks falling into the hands of less abler men.[7]

Events and cultural norms in England over the span of two centuries certainly played the greater part. The Chantries Acts of the 1540s effectively stripped many English schools of their funding by unintentionally diverting these monies into private hands.[8] In 1570, a series of statutes drafted by Queen Elizabeth's administration formally excluded mathematics from university curricula in the belief that the subject belonged in the province of technical education.[9] Since arithmetic was viewed by the English elite as a "mere" trade, it was taught as one, with the textbooks essentially serving the role of reference handbooks as opposed to being genuine guides for instructing the novice.

If this wasn't enough, then came the 1600s with the turmoil of the Thirty Years' War followed by the turmoil of the English Civil War, followed by the turmoil of the Restoration. These events were to all have such detrimental effects on mathematical education in England that it would take the Royal Navy's newly minted Mathematical School at Christ's Hospital (1673) more than thirty years (with the exception of a brief two-year interlude) to find a competent teacher.[10] Even more startling, from the vantage point of today, is that the immensely talented Samuel Pepys (a graduate of Cambridge at age twenty-one), didn't learn how to multiply using a times table until he was twenty-nine years old![11] Two decades after this, Pepys's name, as president of the Royal Society, was to appear on the title page of the most influential scientific document of all time, Newton's *Principia* (1687).[12]

In a very real sense, a set of political, social, and cultural events and norms converged in England during the late Renaissance to create the "perfect storm" against an effective mathematics educational system. Moreover, like the Great Red Spot on Jupiter, this storm was resilient: able to last for several hundred years.

But an ineffective system doesn't mean a 100 percent broken one, for in the middle of it all, the technical brilliance of men such as Thomas Harriot, Christopher Wren, Newton, Edmund Halley, Joseph Priestly, Charles Hutton, and many others still shone forth.

But not as often as it maybe should, for it is likely that the existence of a countrywide system of poor mathematics education (with its default and haphazard reliance on tutors outside of the school to help make up the difference) makes the appearance of such mature and developed mathematical and scientific talent a much greater victim of chance than it naturally already is.

In any event, this ineffectual system with its abstruse textbooks was the one that was eventually imported to the fledgling American colonies. Some of the more well-known English texts of this era are listed here:

- *Hodder's Arithmetic* (1661) by James Hodder: First printed in London in 1661, the twenty-fifth edition was published in Boston in 1719, thus becoming the first-known text on arithmetic to be printed in the thirteen colonies.[13]
- *Cocker's Arithmetic* (1677) by Edward Cocker, published posthumously by John Hawkins: The preeminent school textbook of this period, it was reprinted in more than one hundred editions. Benjamin Franklin, after having failed twice at arithmetic in school, read this book at age sixteen and claimed to have mastered it completely:
  "And now it was that being on some Occasion made asham'd of my Ignorance in Figures, which I had twice fail'd in learning when at School, I took Cocker's Book of Arithmetic, and went thro' the whole by myself with great Ease."[14]
- *Schoolmaster's Assistant* (1743) by Thomas Dilworth: Dilworth's book was endorsed by more than fifty English schoolmasters and used by schools in Boston during the American Revolution. In this text, Dilworth advocates strongly for the education of girls.[15]

Arithmetic Education in America: Eighteenth Century

In the American colonies, instruction was given via imported texts from England. The colonies, of course, initially languished behind the mother country and it wasn't until 1729 that Isaac Greenwood wrote and published the first arithmetic text authored by an American. Titled *Arithmetick Vulgar and Decimal: With the Application Thereof, to a Variety of Cases in Trade and Commerce,* the work was published in Boston by S. Kneeland and T. Green.[16] It was in no way able to challenge the supremacy or replace the use of English textbooks in the colonial schoolroom.[17] Such was the state of math education in the country, that it wasn't until 1780 that elementary arithmetic was moved down from the senior course of study at Harvard University to the freshman year.[18]

After the American Revolution, as the fledgling democratic nation sought its own identity, education of more of its citizens came to be viewed as important.[19] It was also viewed as important that the new country have school books authored by Americans—the future cliché "made in the U.S.A." being evident here. The first such book on arithmetic was authored by Nicholas Pike in 1788 and titled, *A New and Complete System of Arithmetick Composed for the Use of the Citizens of the United States.*

Upon receiving a personal copy of the book, George Washington wrote a letter to Pike on June 20, 1788, in which he stated: "It merits being established by the approbation of competent judges, I flatter myself that the idea of its being an American production and the first of the kind which has appeared, will induce every patriotic and liberal character to give it all the countenance and patronage in his power."[20] (See complete letter in Appendix B.)

Since most American schoolchildren could not afford arithmetic text-books, the classroom drama played out with the instructor verbally stating the rules to the class, using either the book or personal notes, while the students copied down what was said on a slate or in a copybook called a cipher book.[21] This method was not as effective with colonial kids as it had been with medieval Italian children of commerce, who in addition to probably having a capable teacher or tutor, also knew exactly why they were learning the material and where it would be used. On the other hand, these eighteenth-century American schoolchildren (who often didn't have a capable teacher), would have undoubtedly found this method of instruction a very difficult way to learn mathematics.

This state of affairs would change some with the introduction of black-boards (from Europe) into the American classroom. With a blackboard, a teacher could now instruct and show an entire class a calculation or procedure all at once. Few classroom innovations can rival this one in importance.

The report of the superintendent of public instruction in Michigan stated in regards to blackboards in the 1840s:

> Now, it is safe to say that no mechanical invention ever effected greater improvements in machinery, no discovery of new agents more signal revolutions in all the departments of science, than the blackboard has effected in the schools; and certain it is, that no apparatus at all comparable with it for simplicity and cheapness, has to such a degree facilitated the means, and augmented the pleasures of primary instruction.[22]

A Call for Reform in Early Nineteenth-Century America

By the early 1800s, the system of having students simply copy down and memorize rules and examples, with little or no understanding, was increasingly being viewed as totally unsatisfactory for the general population of youths. Benjamin Latrobe, a prominent Philadelphia architect and engineer, stated in 1798, "Arithmetic is generally a heavy study to boys, because it is rendered entirely a business of memory, no reasons being assigned for the rules. A schoolbook of arithmetic accompanied with demonstrations is much wanted."[23]

In 1818, Samuel Goodrich wrote:

> Arithmetic is an abstract, and at first a very uninviting study. Children who do not look to its utility, can see nothing in it but an unmeaning change of figures. . . . It seems, therefore, every effort should be made to clothe this study in somewhat attractive colors; Our school arithmetics are written in a rigidly technical style, which . . . is utterly incomprehensible to children.[24]

During the time from 1800 to 1820, a good number of new American textbooks made bold claims that they were the ones to improve the situation. Not a single one of them went far enough, however. Not until 1821, that is, when Harvard graduate Warren Colburn of Massachusetts wrote a book that finally freed itself completely from the stranglehold that the rules-based, non-demonstrative commercial textbooks had held on education in this country for nearly two centuries.

To do so, however, required that he employ the same methods and ideas being used far across the Atlantic in Switzerland by a fatherly old man—who was destined more than any other single educator to affect the direction and structure of elementary education in the West, and who was in 1821 living out the last few years of his spectacularly influential life.

Whose Fault Is It?

Who was to blame for the deficiencies in elementary arithmetic education? One can take the easy route and simply blame the textbook writers and leave it at that. Clearly, the books they wrote were inadequate for teaching a general population of children even up to fifteen and sixteen years of age.

But is this really fair?

Textbook writers had to write books that publishers wanted. Publishers had to print textbooks that schools and towns would buy. Schools and towns wanted to hire competent teachers. How to demonstrate such competence? Certainly, by showing (or at least feigning) familiarity with the standard methods of instruction using the standard textbooks! This didn't leave much room to change things.

Too simplistic an account? Absolutely, but the educational system of a country is a complex, highly interactive system where events in one area can dramatically affect the entire system. If such a system is deficient it is usually due to a web of complex interactions not simply just one broken component, such as inadequate textbooks.

And what do we even mean when we say the system was deficient? Clearly, by the 1700s a lot more individuals were familiar with arithmetic than were in, say, 1300 CE. In terms of numbers of people, much progress had clearly been made. But the world was becoming more complicated too. This, combined with the continually emerging idea of education for all, placed ever-increasing demands on the system. The system was seen to be deficient in part because it was having difficulty keeping pace with these new demands.

In England, we saw that a complex of reasons were probably responsible for the struggles. What about the other countries of Europe? Was formal education better there? Not really, if we take the writings of reform-minded educators of the 1700s as a reliable guide. It would appear that formal education in general, and certainly in arithmetic, was severely deficient throughout the whole of the continent as well.

But promise lay on the horizon. In fact, conceptual jewels to bankroll that promise had lain strewn on the ground like glowing embers from as early as the 1500s, if not before. But unfortunately, the best of these ideas had yet to be used in any systematic, large-scale way.

By the mid-to-late eighteenth century, however, the Industrial Revolution was in full swing and conditions were finally ripe for change. This time, the thoughts and actions of many educational reformers would finally take root and lead to massive change on a grand scale. It was to be a truly extraordinary time in the history of education, and one of the big winners to come out of it all would be elementary arithmetic. No longer would it be relegated to the specialist school, to the rare self-taught individual, to those of means who could afford a private tutor, or to the senior year of study in elite colleges. For better or worse, it would take center stage in elementary education coming to rival even reading, writing, and spelling in importance.

Let's take a glimpse at the energy, excitement, and brilliance of this time for education. Surely it must tell us a little about ourselves and our present systems of instruction.

Education Reform on the European Mainland

John Amos Comenius

So far we have neglected to mention the scientific revolution—one of the most important paradigm shifts in human history. Surely, it must have had some effect on elementary mathematics education. The answer is that it did, but not in as direct a manner as some might expect; for the great new and influential ideas on what should happen in arithmetic education, as well as with basic education in general, came not from the "full-time" mathematicians and scientists but rather from the generalists—educators and philosophers, men such as John Amos Comenius, Jean-Jacques Rousseau, and Johann Heinrich Pestalozzi.

Francis Bacon (1561–1626), the great prophet of the scientific revolution, summed up the mood when he stated: "Men have sought to make a world from their own conception . . . but if, instead of doing so, they had consulted experience and observation, they would have the facts and not opinions to reason about, and might have ultimately arrived at the knowledge of the laws which govern the material world."[25]

A new breed of scientist, including the likes of Galileo Galilei, Johannes Kepler, Evangelista Torricelli, Newton, and Antoine Lavoisier agreed with Bacon and would begin to consult that experience and observation, and then fuse their findings together in writing with mathematics to usher in what has appropriately been called a revolution in the scientific enterprise.

Scientists were not the only ones who began to apply this "new" way of thinking. Visionaries such as the Czech John Amos Comenius (1592–1670) applied these ideas to the problems of education. And in his case, we know he looked directly to Bacon for his inspiration—believing that "we should look to nature to find out how knowledge takes root in young minds."[26]

Comenius also subscribed to the idea of universal education for all and extended that idea to include girls as well as boys. He is sometimes called the father of modern education. He achieved international acclaim in 1631 after publishing his *Janua linguarum reserata* (The Door/Gate of Languages Unlocked), which demonstrated an interesting, new way to teach Latin.

His fame was such that people from other nations sought his advice on education. The Swedish government made him an offer, in 1638, to help reform their school system, and at the request of Parliament, he went to England in 1641 to do that exact thing—working closely with the brilliant and encyclopedic Samuel Hartlib. He was forced to leave a year later with the outbreak of the English Civil War—at which time he answered the call from Sweden.[27]

Sometime during this period, he is said to have been offered the presidency of America's newly minted Harvard University (although some have questioned this).[28]

In 1657, he authored the children's book, *Orbis Pictus Sensualium* (The Visible World) which is generally considered the first illustrated education book written for children. In this book, he attempts to teach Latin in a manner that is similar in spirit to the present-day methods employed by some software systems to teach foreign languages.

His theories on teaching are best espoused in *The Great Didactic* (ca. 1630s). In this work, he anticipates future reformers in several areas. Unfortunately over time, and with the destructive continental wars that occurred during the prime of his life, much of his innovative thinking was forgotten or credited to others. His reputation experienced a restoration of sorts in the late 1800s when the idea of universal education actually became a reality in many countries, and people began to realize just how far in advance of his era was this Czech genius.

Jean-Jacques Rousseau

Another vigorous proponent of reform in the educational practices of the times was the Swiss philosopher Jean-Jacques Rousseau (1712–1778). Like Comenius, he believed that educators should stop imposing their own ideas and designs without observation and look to nature for guidance in how best to instruct. He directly addresses the question of how children learn in his great work, *Émile* (1762):

> We do not understand childhood. . . . Our teachers will always be seeking the future man in the child, instead of attempting to understand the child as he is and before he becomes a man . . . this is the question I have set myself to study. . . . Childhood has its own way of seeing, thinking, and feeling, suitable to its condition.[29]

Rousseau believed that both the classroom environment and the textbooks themselves should be designed to fit the world of children, rather than the other way around of compelling children, by force if necessary, to conform and fit to the existing modes of instruction. In his commentary on Rousseau, author Robert Quick states of the philosopher's era, "Children have been treated as if they were made for their school books, not their school books for them."[30]

The reaction to much of Rousseau's work was not as he wanted. His *Émile* contained statements that offended the Church and the book was symbolically and publicly burnt in 1762, forcing the philosopher to flee from Paris.[31] He was also the target of ad hominem attacks from others including his eminent contemporary Voltaire, who accused Rousseau of hypocrisy in writing *Émile* given the way he treated his own children. In reference to his writings, Voltaire wrote to Rousseau in 1761 that "One feels like crawling on all fours after reading your work."[32]

In spite of his many professional problems, however, Rousseau's works were to be influential. In particular, they were to take root in the work of a younger countryman who, along with his able associates, was destined to set the world of elementary education in Europe on fire. His name was Johann Pestalozzi.

Education Reform Incarnate

Johann Heinrich Pestalozzi (1746–1827) was born in Zurich, Switzerland, and during his lifetime became the living embodiment of the words uttered below by two very famous men:

> Our greatest glory is not in never falling, but in rising every time we fall —Confucius.[33]

> Imagination is more important than knowledge—Albert Einstein.[34]

Pestalozzi failed disastrously as a pastor in his first attempts at running a service (forgetting some of the words to the Lord's prayer), didn't complete his legal studies, then failed as a farmer, then failed yet again when he turned his hand at cotton spinning and also failed several times as a schoolmaster.[35] He was a poor administrator, was spurned by Napoleon, saw his educational scheme rejected by the French mathematician Gaspard Monge, had his vision dismissed by the famous English educator Andrew Bell, and was thought an eccentric dreamer at best by many of his fellow Swiss.[36]

According to the nineteenth-century Swiss historian Charles Monnard, "He had everything against him; thick, indistinct speech, bad writing, ignorance of drawing, scorn of grammatical learning. . . . He was conversant with the ordinary numerical operations, but he would have had difficulty to get through a really long sum in multiplication or division."[37] In fact, with the exception of his marriage, nearly every big undertaking that he attempted ultimately failed within his own lifetime.

And yet in spite of it all, this dreamer scored spectacular success upon spectacular success especially in his experimental schools at Burgdorf and Yverdon, Switzerland; successes of such magnitude that they probably vaulted him into the position of the single most influential educator in the post Renaissance west.

How did it happen?

Undoubtedly the force of his imagination and ideas as well as timing contributed to his success, but it was also the joining of forces with very capable associates that made his achievements possible. These associates would help him to transform his ideas and theories into dramatic instructional action. Since these men were so integral to his success as he was to theirs, some of the more well-known ones we will mention: Hermann Krusi, Johannes Niederer, Gustav Tobler, Johannes Buss, and Joseph Schmid.

So what was so special about this Pestalozzi?

One would not be too far off in calling him the zealous prophet of educational reform—the fiery incarnation of the idea; for Pestalozzi spoke with a metaphorical fire both against the existing system of education and for his grand vision of a new universal system of education based on psychology. In his book *Gertrude Teaches Her Children* (1801) he rails:

> But for a moment picture to yourself the horror of this murder. We leave children up to their fifth year in the full enjoyment of nature. . . . And after they have enjoyed this happiness of sensuous life for five whole years, we make all nature round them vanish from before their eyes; tyrannically stop the delightful course of their unrestrained freedom; pen them up like sheep, whole flocks huddled together, in stinking rooms; pitilessly chain them for hours, days, weeks, months, years.[38]

> Friend, this view of things led me naturally to the conviction that it is essential and urgent, not merely to plaster over the school-evils which enervate the great majority of the men of Europe, but to heal them at the root—that consequently half-measures in this matter will easily turn into second doses of poison, which not only cannot stop the effects of the first, but must surely double them.[39]

Pestalozzi thought big. His bold vision was for a universal method of instruction that encompassed all disciplines. If applied properly, he felt that this universal method could make average teachers as effective as excellent and inspirational instructors.

In a sense, he wanted to capture the genius in excellent teaching (the genius that inspires and engages students to really learn), bottle it up, if you will, and then dispense it freely to all would-be instructors. Doing so would benefit all students, especially the poor, allowing them to lift their station in life with a consequent benefit to the greater society at-large.

But Pestalozzi was not just a dreamer, he could be very practical as well. He knew that he could not accomplish this task alone and needed the help of other capable men who were as passionate about change as he was, and in his mid-fifties he found them. Together with these men, Pestalozzi's plan was to study and experiment to learn the true laws of education.

Pestalozzi agreed with Comenius and Rousseau that one should look to nature for the answers on how to educate. He felt that children were already learning very naturally before they ever went to school and that school should be a continuation of the methods that worked to educate the child in these earliest of years. This involved looking at children to see how they actually learned. Pestalozzi famously said that he wanted to "psychologize instruction."

What were his ideas on elementary arithmetic education?

Pestalozzi wanted to tear the guts out of the rules-based arithmetic textbooks as well as the methods employed in using them. He felt that the teaching of arithmetic from first principles using the HA numerals was severely at odds with nature and consequently was wrong. For him, learning the rules of arithmetic with numerals should occur at a much later stage of the game after his famous "*Anschauung*" had taken place. We again turn to his words to glimpse his thoughts on arithmetic:

> How many times is seven contained in sixty-three? The child has no real background for his answer, and must, with great trouble, dig it out of his memory. Now, by the plan of putting nine times seven objects before his eyes, and letting him count them as nine sevens standing together, he has not to think any more about this question; he knows from what he has already learnt, although he is asked for the first time, that seven is contained nine times in sixty-three. So it is in other departments of the method.[40]

Pestalozzi believed, like many reformers, that concrete understanding in the child must necessarily precede abstract or symbolic understanding, and that instruction should always proceed with this in mind.

He felt that this harmonized the best with nature. He felt that if one focused on the conceptual as opposed to the symbolic, arithmetic could be taught front and center at a much earlier stage—from the moment that children entered school around six as opposed to waiting until they were eleven or twelve. He, like Comenius, also strongly believed in the use of picture books as an aid in understanding words and of manipulatives in understanding arithmetic (both quite modern ideas).

He even gave a name, "*Anschauung*," for the process, in the student, by which the concrete and vague notions of "a thing" began to crystallize into a firm, unshakeable idea or concept.[41] At this magical point he felt the mind would be ripe to receive a symbolic description in the form of a name, word, or numeral for the idea.

Fame and Influence

Pestalozzi was already a very famous man before he began his landmark work in the Swiss schools at Burgdorf and Yverdon. He had already become something of a celebrity twenty years earlier due to his work as a schoolmaster and the publishing of his book, *Leonard and Gertrude*, in 1781.

This book was a novel describing rural life in Switzerland. Written with a much deeper meaning in mind than simply romanticizing life in the Swiss countryside, that meaning was lost on most readers, and the book was simply read as a good and entertaining novel. This very much bothered Pestalozzi, and over time he authored several sequels which were designed to better expound his educational ideas but none of these books enjoyed a similar popularity with the general public.

His fame was already such that on August 26, 1792, the French National Assembly decreed honorary French citizenship upon him and seventeen other eminent men: citing these eighteen , "as men who in various countries have brought reason to its present maturity."[42] Among the honorees were Pestalozzi; American founding fathers George Washington, James Madison, and Alexander Hamilton; author of *Common Sense* Thomas Paine; discoverer of oxygen Joseph Priestly; British abolitionists William Wilberforce and Thomas Clarkson; German educator Joachim Campe; and hero of America and Poland General Thaddeus Kosciuszko.[43]

His work in the early 1800s, however, had the most far-reaching impact. Men from all over Europe, some at the request of national governments, were sent to Pestalozzi's institute to observe. His influence was the greatest of all in German-speaking Prussia.

In October 1806, Prussia suffered a series of crushing defeats at the hands of Napoleon Bonaparte in the twin battles of Jena and Auerstedt (Auerstädt). The terms of her eventual surrender the following year were harsh and humiliating. Prussia was forced by Napoleon to fire her foreign minister and her king was personally disgraced by the dictator at the negotiating table. In addition, Prussia was stripped of half her land, half her population, and her richest provinces.[44] A "deep depression" descended upon the people and many vowed never again.[45]

In response, Prussia was to totally remake and fashion herself in a way, on a scale, and with a speed rarely seen in nations. Military reformers such as Gerhard Scharnhorst, August Gneisenau, and Carl von Clausewitz, in an attempt to institutionalize military excellence rather than rely on the rare military genius were to revamp the Prussian military and make it the envy of the world.[46] Education reformers sought to accomplish the same with the Prussian school system.

The leaders in Prussia decided that the way Prussia would lift itself to greatness was through the education of its people. King of Prussia, Frederick William III, said, "We have lost in territory, in power, and in splendor; but what we have lost abroad we must endeavour to make up for at home, and hence my chief desire is that the very greatest attention be paid to the instruction of the people."[47] When it came to elementary education in Europe at this time the main event, the "fountainhead" as one of the king's ministers called him, was Pestalozzi; so Prussia was naturally led to him.

Over the years many young Prussians went to Switzerland not just to merely observe but to be totally immersed in Pestalozzi's methods. Among those who visited were the philosopher Johann Fichte, the influential pedagogist/philosopher Johann Herbart, geographer Karl Ritter, education historian Karl von Raumer, and most notably of all, Friedrich Froebel, the founder of kindergarten and whose contributions and influence on childhood education would come to rival those of Pestalozzi himself.[48]

It was by such sharing of information that the meat in Pestalozzi's methods was absorbed into education throughout Prussia. Already as early as the 1760s, Frederick II (the Great) had instituted compulsory education for all citizens thus bringing to fruition, in that country at least, Comenius's idea of universal education. Many countries, the United States in particular (with Horace Mann leading the way in the mid-1800s), would slowly follow Prussia's lead in this regard as well as in some of its reforms. Thus Pestalozzi, by way of Prussia, had a very powerful influence on education in the United States.

Pestalozzi's influence on elementary education in the United States was not limited to the Prussian viaduct. Several others introduced Pestalozzi earlier and more directly into this country. Joseph Neef, a direct associate of Pestalozzi's at Yverdon, opened the first American school modeled on the principles of Pestalozzi in Philadelphia in 1809.[49]

Other influential proponents in New England soon followed suit, introducing the Swiss reformer's ideas in one shape or another into the public consciousness via reform methods in existing schools, the founding of normal schools for teacher training, or simply raising a greater awareness to pay attention to public education. [50]

Edward A. Sheldon founded the Oswego Primary Teacher's Training School (which would eventually grow into the prominent and influential Oswego State Normal and Training School) in upstate New York in 1861. Sheldon wanted the school to teach the Pestalozzian ideas of object training so he hired Pestalozzi expert Margaret Jones as the head teacher. The impact of this institution on education in the United States was significant as teachers trained here fanned out throughout the country spreading the methods with them.[51] Many other normal schools in kind modeled themselves along the lines of the Oswego methods. The influence of this school was felt as far away as Japan.[52]

Quite a number of books have been written on Pestalozzi and his influence. Hopefully, this section has captured some of the excitement surrounding him and his ideas, but he did not exist in a vacuum. Other great educators and philosophers such as John Locke, Johann Basedow, Immanuel Kant, Ernst Trapp, Joseph Jacotot, and J. F. Oberlin also made major contributions to educational practice and thought during the Enlightenment. Some such as Vives and Comenius actually anticipate him by centuries in many respects. However, it was Pestalozzi, more than any other who, in regards to elementary education, "gave form and life to the vague aspirations of his age."[53]

His ability to inspire other men to action is the stuff of legend. One man, John Synge (1788–1845), visited Pestalozzi's institute almost by accident (and only after much coaxing by an acquaintance). He had intended his visit to last for only a quick hour or two, but he was so impressed by what he saw going on that he ended up staying for three months.[54] On his return to Ireland, he set up a local Pestalozzian school of his own on his own land.[55]

Perhaps the most telling feature of his pedagogical magnetism was that even those who disagreed with him on some points of his methods, including Raumer, Herbart, Fichte, and Froebel, were nevertheless still profoundly inspired by him.

A testament to his enduring popularity in Switzerland is a survey conducted in 2008 by the Swiss newspaper *SonntagsZeitung* in which a thousand citizens were asked to list the fifteen most significant Swiss of all time. In the rankings, Pestalozzi was listed fourth (the young immigrant Albert Einstein was first). The complete list may be seen in Appendix C.

There are a number of ways one can spread knowledge of reform ideas, and a few of the major ones certainly include the following:

- Starting your own school and having others observe what you do and then mimic it to some degree by starting their own schools.
- Training teachers according to the reform methods and having the word spread as they obtain employment in diverse places.
- Writing textbooks and curricular materials based on the reform methods and disseminating these in large numbers.

Pestalozzi's reform efforts were able to spread via all three of these avenues. We have already discussed the first two in some detail, now we turn to the last. If you recall, before discussing education reform in Europe and its spread, we broke off from discussing the dismal state of arithmetic education and the accompanying texts in the United States in the early 1800s.

We now return to New Englander Warren Colburn and the sensational book he authored on elementary arithmetic. The success of this text along with its subsequent editions emphatically showed that the fires of Pestalozzian reform in Europe were more than capable of leaping across the Atlantic to burn brightly in the New World.

Arithmetic on the Plan of Pestalozzi

In 1821, Warren Colburn (1793–1833), published a book entitled *An Arithmetic on the Plan of Pestalozzi, with Some Improvements*. It (and its subsequent editions as *First Lessons in Arithmetic on the Plan of Pestalozzi*) represented a radical departure from the American textbooks used before it (see Appendix D). In addition to doing so in the title of the book, Colburn acknowledged the system of Pestalozzi's contribution to his work, in his preface, but also emphasized that the examples used in the textbook were of his own making.[56] The impact of this textbook on elementary arithmetic education in the United States was electrifying. In just six short years from its printing, its author was elected to the American Academy of Arts and Sciences—an organization which only five years prior had elected, as foreign members, two of the greatest mathematicians of all time: Carl Friedrich Gauss and Pierre Simon Laplace.[57]

The timing was such that the book was immediately used in many schools.[58] It is likely the most influential elementary arithmetic textbook in American history. By 1856 more than two million total copies of all editions had been sold, with about 100,000 and 50,000 copies being sold per year, respectively, in the United States and Britain.[59] If true, in today's numbers, this would amount to somewhere between 10 and 20 million total copies being sold within the first 35 years of printing with about 1 million and 120,000 copies being sold per year, respectively, in the United States and Britain. By 1890, an additional 1.5 million copies were said to have been sold.[60] It was also translated into other languages as well. Such was the enduring popularity of this book, that at least seven editions were eventually printed in the Hawaiian Islands before all was said and done.[61]

Compare the first exercises in Dilworth's book (Appendix A) with the first exercises in Colburn's—a vast, vast far-reaching difference. The differences were so sweeping, in fact, that they aided in the total restructuring of elementary arithmetic education in this country.

Dilworth's book, from the start, works with abstract HA numerals that represent large quantities—numbering in the hundreds of millions in some cases. These quantities are so vast and the symbols so abstract that they lie out of the realm of conception for most students who were encountering them for the first time (remember this is the system that, in his very own words, twice defeated the brilliant Benjamin Franklin).

Colburn's book, on the other hand, starts with small values that the student can visualize. Moreover, he begins with familiar words and concrete objects that children were very comfortable with as opposed to abstract HA symbols. His initial aim, like Pestalozzi's, was to give students a thorough grounding in the concepts (in an effort to achieve *Anschauung*) before moving on to performing calculations with figures (introduced in the book later).

The simplicity of Colburn's text meant that children no longer had to wait until the age of eleven or twelve to learn arithmetic but could be exposed in a systematic way to arithmetic much sooner—as soon, in fact, as they entered school. Children could be conceptually taught many of the fundamentals of arithmetic before they even learned to read or write. Affecting this change could be as straightforward as a teacher of six-year-olds simply deciding to use the first portions of the book.

Thus, in a short span of time, the first stages of elementary arithmetic education in schools, where Colburn-like texts were used, could leapfrog down five or six years! This same result awed visitors of Pestalozzi's schools.

Moving the age of first contact down to the early grades led to females in large numbers being exposed to arithmetic for the first time. In early America, the majority of girls fortunate enough to receive an education were done with their formal schooling before age eleven or twelve and thus never had this opportunity. But no longer was this the case, and girls in ever-increasing numbers began to acquire facility in arithmetic.

Regardless, female performance in arithmetic, as well as in other areas, came under particular scrutiny by many educators, as mathematics was associated with rational abstract thinking and many questioned if it was a good thing (or even if it was possible in general) for "more emotional" females to waste time studying subjects, when they needed to learn how to be good homemakers.[62] Of what use would a female be who knew grammar and mensuration but could not bake a loaf of bread?[63] One writer states, "I do not trifle. To be poisoned is a serious matter: and poisoned that man is sure to be, and his children too, whose wife is . . . unskilled in the culinary art."[64]

Strong opinions regarding the capabilities of females to learn logical subjects persist, in some corners, to this very day. In spite of such thoughts in mid-nineteenth century America, it ultimately became a moot point. The vast increase in the number of common schools in the middle of nineteenth-century America required large numbers of new teachers, and some of these would have to be women—the majority of men were already employed elsewhere and these employers were to eventually include the Union and Confederate armies.

Economic circumstances were to soon dictate that *preference* actually be given to female teachers since women could be hired at a fraction of the pay of men. Thus if women were going to be teachers, then they certainly would have to learn and master some mathematics and that was simply that.[65]

Nineteenth-Century Math Wars

Colburn's methods while influential and initially popular did not survive unscathed. In time a strong and inevitable backlash occurred and was so effective that some twentieth-century authors who later wrote on Pestalozzi's contributions to elementary education in America fail to mention the name of Warren Colburn at all (undoubtedly because they probably had never even heard of him).

The two main features of Colburn's text were mental and inductive arithmetic. The mental arithmetic aspects can clearly be seen even from the small excerpt given in Appendix D. This led to oral recitation in arithmetic as the values were small enough to do this. Children engaged in oral demonstrations of arithmetic facts just as they engaged in spelling bees.[66] This was one of the key features that made Colburn's texts so popular.

Colburn's inductive arithmetic had a lot in common with today's discovery method. In its most extreme form, the goals of this method are that students will eventually discover (from seeing enough concrete examples) all of the abstract rules of arithmetic (one of the major champions of this method was the French educator Jacotot [1770–1840] who applied it toward language acquisition). A tenet of the method is that if students discover the rules for themselves in this fashion then they will understand them better. In this arena Colburn probably went too far (as did Pestalozzi) and was the most susceptible to attack.

One author in 1834 states, "We have heard some parents of late, expressing a desire that their children might learn arithmetic in the good old way, of rules and examples."[67] In 1839, in an article in the *American Annals of Education*, the writer states:

> There is now an evident tendency to a return to the old mode in which the various parts out of which the great system is constructed are taught in detail, directions are taken upon trust, the memory is employed to fix them, practice is resorted to make them familiar and at last the system as a whole is seen and understood at the end by the combination of elements and parts slowly and somewhat dogmatically communicated.[68]

In his article *Three Absurdities of Certain Modern Theories of Education* (1851), Tayler Lewis clearly thinks that expecting students to discover rules in the manner that Pythagoras, Euclid, or Descartes did is ludicrous.[69]

Even in New England, Colburn's methods never completely won the day, but they still were able to spread throughout the country via his own textbooks as well as similar ones from other authors who subscribed to the same views or were looking to cash in on his success. However, by the 1850s, the attacks on his methods in respected journals had been successful enough that the thinking had tilted back in favor of the older methods.[70]

Regardless of its critics, this reform spearheaded by Colburn was the first to really shake up the existing infrastructure of instruction in elementary arithmetic in America—an imported infrastructure that, by the time of the printing of *An Arithmetic,* had been in place in the country for more than a century and a half. And although many of his ideas were eventually held in check, Colburn's efforts were nevertheless highly influential and long lasting— especially in the following ways:

- They greatly assisted the move to push elementary arithmetic instruction in America down to the earliest grades.
- They helped make instruction in basic arithmetic available to thousands of American girls. This was instrumental in transforming the entire face of elementary education in the United States.
- They had irreversible effects on elementary arithmetic textbooks and instruction, for the textbooks after him that claimed to be a return to the old ways were still far different from the textbooks before 1800. They were a mix of the old with the new. They included the old, with the rote memorization of rules and calculations using the HA numerals, mixed in with the new, which involved more examples, more practice, more visualization, and more logical sequencing of material from simple to difficult (leading to the abandonment of the one-size-fits-all textbooks and the introduction of graded texts for students of different ages), all on the initial backdrop of much smaller numbers. This new manner in these textbooks would eventually come under attack by future reformers. It is important to remember, however, that this new manner was a monumentally positive reform of earlier truly nightmarish methods.

This nineteenth-century reform in textbooks was in a sense America's real transition from the elementary arithmetic textbooks of old, born out of the thirteenth-century Italian merchant schools, to more modern textbooks. Such transitions occurred independently in other countries throughout the nineteenth century as well. These transitions occurred out of necessity, with the influx of ever-increasing numbers of students with varying backgrounds and goals, and out of a recognition that better methods of instruction needed to be found to educate the general population of young people.

In describing Colburn's contributions to American education, historian Patricia Cline Cohen states, "the vast diffusion of numerical skills from the 1820s to 1900 owed much to his influence."[71]

If we jump to the present for a moment, the average American today clearly has far greater quantitative literacy than the average American had in 1800 (e.g., by being able to do such simple things as tell time from a clock, build a personal budget, and understand a weather report) which indicates that, in spite of the serious issues that are seen to still exist now in mathematics education, truly amazing strides have been made forward. Indeed, some of the very beautiful dreams entertained by the more enlightened reformers of the past have in some measure been realized in this country (and even more so in many other countries), and Warren Colburn was central in making the first major steps toward these realizations in the United States.

Conclusion

We see that the story of elementary arithmetic education is large and intimately connected with the societies it serves. In two prose-filled chapters, we have clawed our way through a skeleton of highlights to barely reach just mid-1800s America. So much of interest has happened since, but it is best to stop now lest we start to feed the fiction that our attempt here is anything more than a brief tour of certain specific highlights through what is an immense topic.

In attempting any snapshot of events that happened so long ago, one is totally dependent on a host of sources from both the past as well as the present, and consequently is susceptible to errors. As such, I freely take full responsibility for any errors in fact, interpretation, or omission that may exist.

Time to summarize! We have focused on the following major story lines:

- The introduction of the HA numeral system into Europe.
- The struggle and eventual triumph of this system over the old ways of reckoning (e.g., the counter abaci-Roman numeral system).
- The early methods and aims of elementary arithmetic instruction involving the HA numeral system (in the classroom and via books).
- The two centuries' long efforts to improve these early methods, with a closer look at some of the reforms put forth by Pestalozzi and Colburn.
  - As mentioned in the introduction to the book, affecting change in education on a massive scale takes the work of many dedicated people. The same is true here. The omission of many names and texts in our discussion of these reforms is in no way meant to detract from them nor give every ounce of the credit to only two men no matter how exceptional they were.
- The inherent difficulties in teaching mathematics to large groups of diverse human beings in a complex society.

As many of you know even from your own personal experience, by the age of ten for thousands of youths, of average mathematical ability, it is now possible to become quite competent in performing all four of the basic operations using HA numerals.

This is an astounding achievement of elementary education!

It represents truly spectacular change indeed: from the days of the two-pronged process of recording in Roman numerals and calculating on the abacus (where a youth in his teens might apprentice with a master for a few years to learn the craft of calculation), from the days where conscientious merchants from other parts of Europe might send their children to Italian merchant schools to learn how to multiply and divide in writing to the days in the eighteenth century where basic computation in HA numerals was taught in the senior year at elite colleges such as Harvard.

From our present-day perspective, the magnitude of this achievement can be likened to educators one day figuring out how to teach basic facility in calculus, not to college students majoring in engineering, but to the average twelve year old.

Yet despite this achievement in elementary education and the many reform efforts in American mathematical instruction that have occurred since this first big one, we still are seen as having a crisis in mathematics education today.

Why is this? Why do so many who go through schools in our country still end up neither understanding mathematics past arithmetic nor liking it much either?

These are not questions with simple answers. Educational systems by their very nature are leviathans; so it is hard for them to quickly respond to change (especially for a system as localized as ours), even if such change is warranted. Moreover, everything is complicated by the fact that there is still much that we just don't know about how to effectively teach mathematics, especially beyond elementary arithmetic. Outside of the basic rudiments, much of what is done in mathematics, at least on the surface, is not natural at all. It is very powerful and important, mind you, but it is not easy for most to understand—more research, hard work, and exposition are required.

Given this reality, the reform efforts and hot disagreements of the past can be expected to continue through the present and into the future. To some degree this is understandable as the stakes are high. Education is how we prepare the next generation to partake in their human heritage and add to it. The great American educational theorist John Dewey surely thought so when he stated, "What nutrition and reproduction are to physiological life, education is to social life."[72]

In the face of what sometimes seems like insurmountable difficulties and eternal disagreement, there is much promise on the horizon. But this promise has to be nurtured. Other hope-filled times for the improvement of American math education (such as the powerful efforts spearheaded by university faculty such as David Eugene Smith, Columbia University, and Jacob William

Albert Young, University of Chicago, during the early 1900s), have failed to reach their true potential due to politics, endless bickering, disrespect, arrogance, and outright stupidity among many of the parties involved.

Why such optimism now?

First, the spectacular advances in the relatively new field of cognitive science are starting to make their presence felt. Cognitive science deals with understanding in fundamental ways how the mind works and acquires knowledge. About as interdisciplinary a field as there is, its major components come from at least six areas: anthropology, computer science, linguistics, neuroscience, philosophy, and psychology.[73] In addition to all of the exciting research that has already been done, we are also seeing the rise of an exciting new subgenre of popular works from scientists in this area (this is not an insignificant thing). So much has happened in the last fifty years or so, that it is being called a revolution in many circles, and just as the scientific revolution came to eventually have spectacular effects on educational thinking in mathematics, so too should the cognitive revolution.

Second, many research mathematicians, who have learned valuable lessons from the failure of the New Math reform movement of the 1960s, are now teaming with mathematical educators and other researchers to carve out their own niche in the cognitive revolution. For example, one goal is to learn what types of knowledge are really needed for a person to be an effective mathematics teacher. Such research should prove useful not only to those who aspire to be teachers but also to the thousands of excellent teachers presently working in our schools.

Researchers are also finding out that simply requiring would-be teachers to take more math classes and become more sensitive people is not enough. Nor is it enough to simply create favorable conditions to entice more of America's best and brightest to go into teaching (although this isn't a bad thing); for as some of you may know from personal experience, a lot of very smart and very nice mathematicians (researchers and educators alike) are not effective as teachers. Despite the fact that it has been much maligned in the past couple of decades, this is an exciting time for mathematical education research.

Third, in terms of an overall comprehensive plan of math education, there are other nations that are simply doing a better job of educating their students than we are right now (though none of their systems are perfect either). This is true whether we use a standardized test as a measure or not. While some of the apparent disparity may perhaps be explained away by the fact these nations separate their students into different career paths earlier than we do and only assess their strongest, most of it cannot be as we find upon closer inspection that some countries just have a structure that makes more sense than ours. And although we cannot copy them exactly due to the different natures of our respective societies, there is still much to learn from them ("the wits and endeavors of the world 'being' so many scattered coals or fire-brands").

What's illustrated in the comparative studies done so far is that coverage alone does not equate to understanding. Some of the nations that seem to be outperforming us in K–12 mathematics actually have teachers who take fewer math classes than our teachers do, cover less material than we do in their curriculum, and spend less time doing it.[74]

The issues regarding too much coverage in the American curriculum was found to be the case even in the early 1900s (this time in the schools themselves), when J. W. A. Young, of the University of Chicago, did a comparative study between the Prussian and American mathematical educational systems.[75] Francesco Sacchini, the Jesuit educator from the 1600s, evidently was of a similar opinion when he reportedly stated in regards to students, "Care rather for their seeing a few things vividly and definitely, than that they should get filled with hazy and confusing notions of many things."[76] Much treasure is available for the mining, in both directions, from such international studies.

Finally, much more material concerning the history of math education (as well as the history of mathematics in general) is becoming available as new groups of dedicated, talented, and culturally respectful researchers spring up around the world. A good contextual understanding and respect for both histories is critically useful to all major discussions where serious change is being proposed. Military schools (and sports coaches too) have long known this about the history of their respective disciplines, and have shown such study to be extremely powerful and relevant when used as a potent analytic tool to aid in the building of new plans and strategies for the future.

In almost any subject, when you start to look at its history you generally find that it is massively more extensive and broader than you think. Mathematics education is no exception. Volumes of written works on educational methods and thought are available in English as well as in many other languages—evidence that a lot of people have already thought and worked hard on the eternal issues involving education.

Lack of historical context often leads educators, mathematicians, politicians, and others to replicate past mistakes. Education, perhaps more than any other arena, provides dramatic and recurring illustration to the famous words attributed to George Santayana: "those who cannot remember the past are condemned to repeat it."[77]

With most ideas in education, the mistakes are ones of excess rather than of being totally off base. It stands to reason that since physiologically we need a balanced diet, we would also require a balanced psychological diet as well. In all likelihood, nice healthy portions, containing the strengths of lots of the methods, are probably what most students need.

As the 2008 report of the National Mathematics Advisory Panel states, the various methods put forward are not necessarily in conflict with each other and the strengths of each should be incorporated into an overall comprehensive strategy producing a healthier mental diet for learners.[78]

Lack of historical context can lead to a romanticizing of the past, where people hearken back to a golden age of math education where children learned mathematics as they were supposed to—an age that we currently should aspire to reach again. Whether such an age has ever existed elsewhere in the world, where large numbers of people were involved, is highly doubtful. One thing is for certain, however, and that is that no such golden age has ever existed in the United States nor in the thirteen colonies preceding it.

And regarding attitudes toward the subject, which in the end turns out to be especially important for the overall health of an educational system, a consistently large portion of Americans have always disliked math, questioned its value to them, and been intensely frustrated with it.

We conclude now with a few quotes from educators, scientists, and philosophers from the past and the present.

Quotes on Education and Mathematics

I threw away my drawing-board and ruler, and burst out in rage against mathematics, because it tortures so cruelly those who love it and are eager for it.—Petrus Ramus, sixteenth century French humanist, philosopher, and educational reformer [79]

A task of the teacher is to cultivate a taste and make it a power in the soul. This is accomplished by the aesthetic revelation of the world through instruction. —Johann Friedrich Herbart (paraphrased), nineteenth century German educational philosopher, one of the founders of scientific pedagogy[80]

Grasping the structure of a subject is understanding it in a way that permits many other things to be related to it meaningfully. To learn structure, in short, is to learn how things are related.—Jerome Bruner, contemporary, influential American cognitive psychologist, author of *The Process of Education*[81]

Shameless ignorance in regard to such serious intellectual conquests as are embodied in the mathematical literature does not represent a normal condition on the part of those interested in the history of the human race. On the contrary, such shamelessness is evidence of the lack of the proper aids to enter this literature.— George Abram Miller, early twentieth century mathematician, former president of the Mathematical Association of America[82]

A most sufficient recommendation of the study of old works to the teacher, is shewing that the difficulties which it is now (I speak to the teacher not the rule-driller) his business to make smooth to the youngest learners, are precisely those which formerly stood in the way of the greatest minds, and sometimes effectually stopped their progress.—Augustus de Morgan, nineteenth-century British mathematician, logician, and educator[83]

Psychologically, the teaching of abstractions first is all wrong. Indeed, a thorough understanding of the concrete must precede the abstract.—Morris Kline, twentieth century math historian and celebrated popularizer of math[84]

To give simplicity of form with depth of thought is one of the qualities of the difficult art of teaching.—Charles Laisant, late nineteenth century French politician, engineer, and math education writer[85]

Despite how commonplace it may seem, teaching is far from simple work. Doing it well requires detailed knowledge of the domain being taught and a great deal of precision and skill in making it learnable.—Deborah Loewenberg Ball, contemporary math educational researcher, member of the National Academy of Education[86]

There is a wise proverb that warns us that "however soon we get up in the morning the sunrise comes never the earlier." A vast amount of instruction is thrown away because the instructors will not wait for the day-break.—Robert Quick, nineteenth-century British educator and writer[87]

The most important single factor influencing learning is what the learner already knows. Ascertain this, and teach him accordingly.—David Ausubel, twentieth century educational psychologist, proponent of the theory of advance organizers[88]

One fancies, indeed, that experiments in education would not be necessary; and that we might judge by the understanding whether any plan would turn out well or ill. But this is a great mistake. Experience shows that often in our experiments we get quite opposite results from what we anticipated. We see, too that since experiments are necessary, it is not in the power of one generation to form a complete plan of education.—Immanuel Kant, eighteenth-century German philosopher of the Enlightenment[89]

A lot of educational and cognitive research can be reduced to this basic principle: People learn by creating their own understanding. But that does not mean they must or even can do it without assistance. Effective teaching facilitates that creation by getting students engaged in thinking deeply about the subject at an appropriate level and then monitoring that thinking and guiding it to be more expert-like. —Carl Wieman, contemporary physicist, Nobel laureate, science educator[90]

One can invent mathematics without knowing much of its history. One can use mathematics without knowing much, if any, of its history. But one cannot have a mature appreciation of mathematics without a substantial knowledge of its history.—Abe Shenitzer, contemporary Polish-born mathematician, educator, Holocaust survivor[91]

If you could lead through testing, the U.S. would lead the world in all education categories. When are people going to understand you don't fatten your lambs by weighing them?— Jonathan Kozol, contemporary education writer, activist, National Book Award laureate[92]

Metaphors are the most primitive, most elusive, and yet amazingly informative objects of analysis. Their special power stems from the fact that they often cross the borders between the spontaneous and the scientific, between the intuitive and the formal. Conveyed through language from one domain to another, they enable conceptual osmosis between everyday and scientific discourses, letting our primary intuition shape scientific ideas and the formal conceptions feed back into the intuition.—Anna Sfard, contemporary math educational researcher, Hans Freudenthal Award laureate[93]

IV

ILLUMINATIONS

There are two kinds of light—the glow that illumines, and the glare that obscures.

—James Thurber, American author,
humorist, and cartoonist[1]

What distinguishes a mathematical model from, say, a poem, a song, a portrait or any other kind of "model," is that the mathematical model is an image or picture of reality painted with logical symbols instead of with words, sounds or watercolors.

—John L. Casti, contemporary mathematician,
complexity theory scientist, author[2]

12

Symbolic Illuminations

Man possesses what we might call symbolic initiative; that is, he can assign symbols to stand for objects or ideas, set up relationships between them, and operate with them on a conceptual level.

—Raymond Wilder, American mathematician,
author of *Evolution of Mathematical Concepts*[1]

The mathematical student needs to use the telescope as well as the microscope, even if the latter instrument is the more important for those who desire to become experts along mathematical lines.

—George Abram Miller, American mathematician, educator,
author of *Historical Introduction to Mathematical Literature*[2]

THE NAMING OF THINGS, phenomena, and ideas by signs has literally helped us to create a "symbolic civilization" out of a vast and often confusing conceptual wilderness. Concepts in their rawest form, though potentially very powerful, can be wild, fluid, unruly even contradictory. Symbols have allowed us to establish some order, stability, and "light," if you will, in this vast, untamed, and uncultivated world of ideas. In this chapter, we come full circle by once again reflecting specifically on how illuminating the process of naming with numbers/numerals can indeed be.

The Glow That Illumines

Language allows us to light up our world. With language we can refer, in significant and impactful ways, to things, actions, and so on, without physically pointing them out or even seeing them. The mere mention of the words "Las Vegas" or "Pacific Ocean" by someone on the radio can conjure up a host of images in the listener. The initial stimuli for these images are not the pictures of these places but rather the verbal words conveyed by the announcer's voice. It is these sounds that are first processed by the brain to ultimately produce the images that come into the minds of the listener.

In a metaphorical sense, we can think of these responses to the audio words as the turning on of a set of mental lights. In fact, when neuroscientists study human cognition they image mental activity, in some cases, literally as lit up regions in the brain.[3]

Consider the following two maps of the United States. Notice how, for certain types of information, the naming in the second figure gives more context, depth, and structure (more light if you will) to the map than does the first figure without the names.

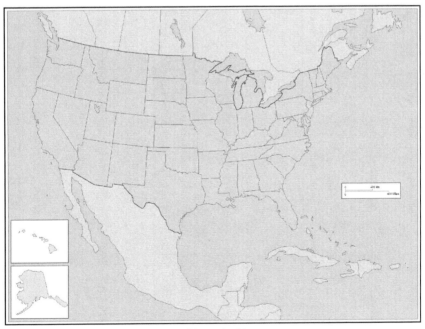

Map of the United States
Produced by the Cartographic Research Laboratory: The University of Alabama

Map of the United States with State Capitals
Produced by the Cartographic Research Laboratory: The University of Alabama

The map with names brings to the fore a host of new questions that would probably never arise from the nameless map, questions such as, which two state capitals are the closest together (Providence and Boston), or what is the southernmost state capital in the continental United States (Austin, Texas). This is the glow that illumines.[4]

You might think that taking the map and pumping it further with more information would bring in even more benefits. But if we do this indiscriminately, we run the risk of adding too much information, leading to a glare that obscures.

The use of language in adding understanding to a subject almost gives it a new dimension.

Quantitative Illumination

In what follows, when we speak of symbolic naming, we mean attaching a symbol or group of symbols (and the ideas they represent) to an object or idea in either a permanent or temporary fashion. This millennia old practice finds one of its fullest expressions when numbers are used in the naming. The ways in which numbers are used as names range from the mundane to the truly spectacular. In chapter 2, we have already discussed some ways that numbers are used as names. Before moving our discussion forward, we summarize these here:

- *Measuring the size of a collection*: If we name each member of a discrete collection of objects by using the ordered sequence of natural numbers {1, 2, 3, 4, . . .} then the last number used in the naming contains complete information about the size of that collection and thus measures it. This is a special type of number naming which we call counting.[5]
- *Ordering collections*: In performing a count of a set of objects, we are temporarily attaching number names to the individual members of that collection. Nothing then prevents us from actually using number names in a more permanent fashion to refer to those objects in place of their proper language names. This allows us to transfer the order present in a number sequence to the collections we name using them.
  - This ordering can be precise, using the consecutive natural number sequence starting at the number one, or less exact, using a consecutive sequence of natural numbers not starting at one, or an ordered sequence of nonconsecutive natural numbers.
  - Examples of using the consecutive natural number sequence include ordering the runners in a race according to how they finish, the pages in a book or the largest metropolitan areas in the United States according to their populations (e.g., largest metropolitan areas in the United States: (1) New York City, (2) Los Angeles, (3) Chicago).
  - Labeling the south facing hotel rooms on the third floor with natural numbers or labeling houses on streets with numbered addresses give examples of using an ordered sequence of nonconsecutive natural numbers.
- *Identification*: If we want to uniquely identify millions of people then naming with numbers has advantages. With proper English language names uniqueness is lost as social conventions lead to many people having the same name, and there is no predictable way as to how such names are chosen.
  - Numbers, on the other hand, are systematic and far outnumber any collection of people that we might care to classify. Examples of number names as unique identifiers include: social security numbers, student IDs, and state driver's license numbers. Other examples of number names as identifiers (not necessary unique) include: credit card numbers, telephone numbers, and IP addresses.

The advantages of naming with numbers are extensive. In the next few sections, we take an interesting cross section of situations involving their use. As always, our hope is that a few insights can be gleaned along the way and that they can be used elsewhere by the reader as conceptual capital.

Navigation into Unknown Territory

If you happen to be totally disoriented or lost, sometimes all it takes is the recognition of a single familiar thing such as a landmark to change everything. The sight of this one thing can have an effect that is worth a thousandfold more than what might appear on the surface. It can allow you to place an entire landscape in context thus giving you the game-changing ability to successfully negotiate your way. Naming with numbers often gives us similar orientations when we are dealing with something unknown:

- *Mileage of a vehicle*: If you are considering the purchase of a seven-year-old vehicle, you care about the total number of miles the vehicle has been driven.
  - What this number does is give you crucial orientation (in the same way that a landmark does for location) as to the history of a vehicle that you may have only seen once or never at all. While there are still many things about the vehicle that you won't know from its odometer reading, knowing its mileage is a crucial component in whether most people would decide to purchase that vehicle or not.
- *The age of a person*: The simple knowledge of the quantity that describes the time a person has been alive can give you crucial insights about a person even though you may not have ever personally met them. For instance, if the person is a ninety-seven-year-old American in 2013, then you can immediately know that they probably have personal memories of the Great Depression, World War II, the Cuban Missile Crisis, the Apollo missions, and Watergate but not of the Civil War, the assassinations of Lincoln, Garfield, or McKinley, or the sinking of the *Titanic*.
  - Those who know their history have a map of events and their approximate dates in their head and these serve as a navigational map of sorts allowing them to place some of the major cultural experiences that a ninety-seven-year-old person may have experienced in the context of that map.
- *Trip navigation*: Navigating on a trip is significantly enhanced when signage and landmarks are available. However, many situations occur where neither signage nor landmarks are available. Ocean travel is one such scenario and numbers come to the rescue in the guise of longitude and latitude. Longitude and latitude assign a set of number coordinates to every location on the surface of the earth. For instance, the spot with number coordinates (latitude 0°, longitude 90° west) is located in the Pacific Ocean on the equator in the eastern portion of the Galapagos Islands archipelago.

- An important component in successful navigation is determining these coordinates—a task extraordinarily easy today but more difficult in the past. When *Carpathia* came to *Titanic*'s rescue, it could do so by being given the longitude and latitude coordinates of the beleaguered vessel.
- Numbers can help us in performing simple navigation while driving, even when road signs are not visible. If we know the distance between town A and town B is 200 miles and that our odometer shows that we have traveled 190 miles since leaving town A, then we know we are about 10 miles from town B regardless of whether or not road signs are visible or hidden by a fog or the whiteout conditions in a blizzard. This information can also tell us that our gas tank may be nearly empty even if the gauge isn't working.
- Simple navigation with numbers occurs when we stay in a hotel. Knowing only the room number is sufficient enough information to find our way with the greatest of ease even though we are stepping off of the elevator onto a floor that we have never seen before in our lives.

More Sophisticated Counting: Measurement

Our ability to visually distinguish between size with precision and accuracy is, as we know, limited to very small numbers. That is, at a glance we can quickly distinguish between a group of two boxes and a group of three boxes. However we cannot on sight quickly distinguish between 998 boxes and 1,000 boxes. The ability to count changes everything.

What counting does is allow us to vastly extend our native abilities to distinguish even minute differences between very large sizes. Counting gives a warehouse manager the ability to conclude that during the night someone possibly stole two of her boxes even though her facility may be covered by a sea of the cardboard containers. She would never be able to tell this on sight alone, yet through counting (tally or order) she can. She could even distinguish the difference between a collection of 10,000 boxes and one with 9,998. Embezzlers beware!

The ability to count allows us to not only say that Hartsfield-Jackson Atlanta International Airport had more passenger traffic than Chicago's O'Hare International Airport in 2010 but to say by how much: 89,331,622 versus 66,774,738, respectively.[6]

Counting allows us to systematically see differences in sizes (large or small) with depth and structure. Visually we can only make crude comparisons for sizes larger than our subitizing ability. That is, we can certainly tell on sight that a group of 207 people is larger than a group of 42 people but we can't say by exactly how much. The ability to count allows us to say by how much with the precise value of 165.

Using an analogy, without counting it's almost as if we can't distinguish individual faces except perhaps when they differ by gender or at least ten years of age but with counting we are able to actually recognize the complete range of faces as we naturally do.

Just as we obtain the ability to "light up" in great systematic detail the crude notions of "more people" or "a larger number of boxes" through the process of counting, we would also like to "light up" other crude notions in great systematic detail as well. Other such crude notions include:

1. Distance: Farther, shorter, longer, taller
   - The distance from Chicago to Los Angeles is farther than the distance from New York to Washington, D.C.
   - Kareem Abdul Jabbar is taller than Magic Johnson.
   - A FedEx cargo van is longer than a Chevy Impala.
   - *Question: Exactly how much farther, shorter, taller, or longer?*
2. Weight: Heavier, lighter
   - The truck is heavier than the car.
   - The book is lighter than the brick.
   - *Question: Exactly how much heavier or lighter?*
3. Time: Longer, shorter
   - The baseball game lasted longer than the tennis match.
   - Alan Shepard's space flight lasted for a shorter time than Yuri Gagarin's.
   - *Question: Exactly how much longer or shorter?*
4. Temperature: Colder, Hotter
   - La Ronge, Saskatchewan, was colder than Bismarck, North Dakota, yesterday.
   - The high in Phoenix was hotter than the high in Alice Springs, Australia.
   - *Question: Exactly how much colder or hotter?*
5. Prices: More expensive
   - A new Hyundai Sonata is more expensive than a new Hyundai Elantra.
   - The silk shirt cost more than the cotton shirt.
   - *Question: Exactly how much more expensive?*

To give structure in depth to the variations in these crude notions, we will need to develop some way to count them. In counting discrete objects such as people, books, and boxes, the criteria for being counted is simply existence as an individual entity. That is, if the object exists in the collection, it is assigned a number or is counted. Each object is a discrete packet all unto itself.

It is not as obvious, however, how to assign numbers to the notions of distance or weight or time since they don't separate themselves into individual packets (i.e., they are not discrete). Can these things really be counted?

In spite of their not being discrete, the trick to devising a way to count these notions is to still define a fundamental discrete unit or quantum for them.

Consider the following line segments:

Segment A

Segment B

Segment B is clearly longer than segment A. However, if we want to state this in a more exact manner not just a crude one, then we need to devise some way to count these segments. We can count both by devising a standard unit by which we can measure them (this amounts to devising our "1", if you will). Let's define our standard to be the following:

One Token

We can use this unit to count both line segments. We do this by counting the number of copies of the fundamental unit it takes to equal the length of each segment:

1   2   3   4   5

Segment A with Tokens

1   2   3   4   5   6   7   8

Segment B with Tokens

The standard unit allows us to now say with more confidence that segment A is 5 tokens long and segment B is 8 tokens long.[7] Thus the crude notion of segment B being longer than segment A can now be replaced with the more exact notion that segment B is 3 tokens longer than segment A.

If someone has a good feel for how long the fundamental unit is then they can have a feel for the lengths that correspond to both 8 tokens and 5 tokens as well as their difference of 3 tokens.

Using the tokens as we previously have quickly becomes a pain in the neck. Let's rearrange them to a more convenient form by using hash marks to indicate each copy of a token. Doing so allows us to craft out a measuring device:

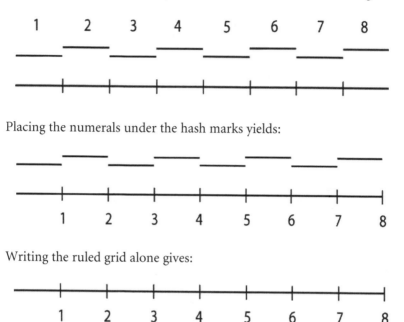

Placing the numerals under the hash marks yields:

Writing the ruled grid alone gives:

Now that the line itself is ruled, we can use it to measure length instead of using repeated copies of the individual tokens. Ruled lines like this are sometimes called number lines and find extensive use across a wide swath of mathematics.

Although we have used a discrete unit to count them, it is important to realize that lengths are still continuous magnitudes not discrete ones. What this means in essence is that when counting a discrete collection, such as a group of people, the size of the group must be one of the natural numbers {1, 2, 3, 4, . . . }. People come in whole units.

With lengths, far more variety exists. We could have situations such as those given below, where a whole number of fundamental units will not fit. The lengths of segments C and D are somewhere between 7 and 8 tokens. There is certainly nothing wrong with this—these scenarios are as valid as the scenarios above. But the quantities 7 or 8 are not exact enough to distinguish between the lengths of C and D or from other lengths that lie between 7 and 8. We need more values (accompanied by the corresponding numeral names). Those values of course are provided if we augment our whole numbers with fractions and more generally decimals.

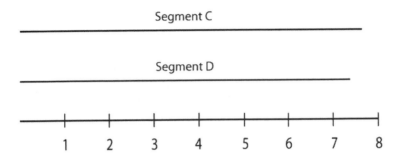

In principle, any place on the line between 7 and 8 could represent the length of some segment. This corresponds to infinitely many more values—or what we call "a continuum of values." With people, you either have 7 or 8 of them, nothing anywhere in between.

We can also count other crude notions such as "heavier" by choosing a fundamental unit or quantum for them as well. For instance, we might choose a certain type of brick as the standard for counting weight and then use a scale to count how many bricks it takes to balance another object. An object requiring 8 bricks to balance it would be said to weigh "8 bricks." Such a system would allow us to give structure in depth and detail to the crude notion of weight just as the *token* allowed for length.

Making precise the crude notion of time by naming it with numbers is one of the most multilayered, fundamental systems of measurement we have. In June, when a meeting for a group of people, one-thousand strong, is scheduled at a hall in Boston for 9 p.m. on December 15, numbers guide their actions.

Without numbers, how will all of the people be able to unambiguously distinguish one night from the other or even certain portions of a given night from other portions? With numbers (namely, the date and the time of day) and the tools of a calendar and clock, these thousand people can with ease synchronize their actions and all show up at the same location at the same time. It doesn't matter if it is dark outside, raining, or even half a year later, as long as they have the required tools, number naming allows them to successfully and collectively navigate through time.[8]

Calculation

One of the most spectacular demonstrations of the power in naming with numbers occurs when, in addition to utilizing their static properties as labels, we turn to their dynamical properties for change through combination with each other. When this happens, the ability of these ideas to transform into other ideas reaches a high art. This high art of transformation is demonstrated symbolically in the guise of performing calculations.

Quantitative ideas are less ambiguous than ideas such as "love" and "hate." We can clearly tell that the collection of letters {A, B, C} has the property of three while the collection of letters {D, E, F, G, H, I} does not. This definitiveness allows us to create a much more precise and special way of communicating quantitatively about things, for in combining the unambiguous ideas of three objects and six objects, we don't get gibberish nor varying interpretations which might be expected but instead get the equally unambiguous idea of nine objects.

And just as there are a multitude of different ways to symbolically describe reality by human languages, a myriad of different ways exist to symbolically describe this quantitative fact:

three plus six equals nine: English

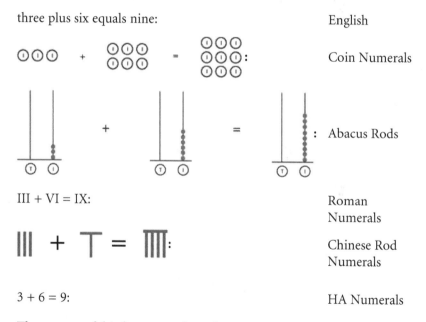

Coin Numerals

Abacus Rods

III + VI = IX: Roman Numerals

Chinese Rod Numerals

3 + 6 = 9: HA Numerals

The content of this fact comes from the way in which the ideas combine not from the symbolic expressions themselves. The symbols simply communicate how the ideas combine in abbreviated form.

This is not the end of the story, however, for the symbols play a far bigger role for us than simply abbreviating or representing an idea. The symbols provide us with a whole new way of looking at what is happening and also provide schematics which permit us to see crucial relationships unscreened.

This allows us to observe patterns in the symbols from which we can build sophisticated algorithms that enable us to obtain (calculate) answers to all sorts of scenarios in a much quicker fashion than we could physically working with the objects alone. We have seen how these patterns can be exploited to create the powerful algorithms we have today for the basic operations of arithmetic.

We can also view the ability to calculate as giving us yet another sophisticated way to "count." Consider the case of the Begay family and their desire to predict how much they can expect to pay for gas on a 1,425-mile road trip. Can they "count" the dollar amount of gas they will use on the trip in the direct way?

Put another way, can they obtain the cost in the same manner that they can, say, count the number of people in a classroom—in a straightforward process of matching off? Unfortunately they can't, as the cost in dollars doesn't relate in a one-to-one fashion with the number of miles driven on the trip. That is, the family can't simply count in a natural way the miles as they drive down the highway and then match these directly with the dollar cost for the gas. In fact, they can't even see the gasoline once it has been pumped into the gas tank.

However, since the total gas cost depends on the components of distance traveled, the average gas price and the gas mileage, its value can be obtained through a calculation. What the calculation allows them to do is to bring these three components into conjunction to generate a numerical dollar value for the total quantity of gas used. Let's assume that the average gas price is $3.00 per gallon and that the vehicle's gas mileage is around 25 miles per gallon. From this information they can obtain what they need as follows:

The number of gallons they will need to purchase for the trip: The gallons come in 25 mile packets so to determine how many of these are needed to cover 1,425 miles, we need to calculate $\frac{1425}{25}$. We use the symbolic algorithm for long division to do this:

$$
\begin{array}{r}
5\ 7 \\
25\overline{)1\ 4\ 2\ 5} \\
^-1\ 2\ 5 \\
\overline{1\ 7\ 5} \\
^-1\ 7\ 5 \\
\overline{0}
\end{array}
$$

Thus a total of 57 gallons are needed.

The cost to purchase 57 gallons at an average cost of $3.00 per gallon: The problem now reduces to the multiplication of 57 × 3. We do this using the standard U.S. multiplication algorithm:

$$\begin{array}{r} 2 \\ 5\ 7 \\ \times\ \ 3 \\ \hline 1\ 7\ 1 \end{array}$$

And we see that the family should expect gas costs to be around $171.00.

The calculation gives a seamless way in which to transform one set of ideas into another set of ideas ("liquefying" them in a sense). Since every quantity or number represents an idea, we start with the ideas 1425, 3, and 25 and through well-defined algorithms (played out on HA number diagrams), we can smoothly and unambiguously transform these to the idea of 171 to obtain the estimated dollar cost of gas for the trip. Moreover, we are able to obtain this estimate on a piece of paper before the family ever embarks on the trip.

If we liken the ability to directly count a quantity to the ability in English to describe a concept or event by a single word, then we can metaphorically liken the "need to calculate" to find a quantity to the need in English to construct sentences to describe a concept. For example, no single English word allows us to describe the situation of "a group of ten people going to Pyramid Lake in Nevada for a picnic on a Sunday afternoon," but this sequence of eighteen words does just fine in describing the event.

And for the record, sentences are capable of describing way more situations than single words can. For instance, the number of physical objects described by the word "house" numbers into the millions but the number of physical situations described by the statement, "the woman deposited money into the bank" easily numbers into the thousands of trillions (quadrillions). This statement is capable of describing all situations involving any living woman (of which there are billions) depositing any amount of money (numbering from one cent, or any other currency, on up) into any bank on the face of the earth (presently more than 7,000 FDIC banks are in operation in the United States alone).[9]

This means that we are not limited symbolically in using numbers to deal only with things that we can directly count. If we cannot directly count something (due to size, logistical difficulties, or inhospitable conditions), we may still be able to perform a calculation to gauge it (i.e., construct a "quantitative sentence" to measure it), and thus, calculation symbolically extends almost in an automatic way the things that we may subsume under the domain of number. Performing strings of calculations using elementary arithmetic gives perhaps the most accessible demonstration of the broad predictive power inherent in mathematics.

Calculations Are Symbolic Events

It can't be emphasized enough that calculations just like direct counts are symbolic events, and in performing them we are speaking or writing about numerical interactions, not controlling them. Also, the calculations themselves, don't directly alter the objects or events being described only what we can say about them, how we think about them, and ultimately what we may ourselves decide to do with them.

The numerals also exhibit the well-defined and reproducible properties of the things they describe, which mean that regardless of who is doing the calculations (if they do them correctly), they should come up with the same answer. This reproducibility is one of the hallmarks of calculation as well as mathematics in general.

Let's look at a few examples to illustrate the discussion.

Symbolic simplifications don't alter objects:

Consider the following coin collections:

We can describe or label, respectively, the number of elements in each collection symbolically as:

Combining collections A and B together yields the collection:

We can describe this symbolically as

Since the objects are all of the same denomination we can simply write 3 + 4 = 7. It is important to remember that this is a symbolic simplification since physically seven distinct pennies are still present. The pennies have not been metallurgically combined to form a single object. They are simply represented as combining to a single object symbolically (Wilder's "symbolic initiative").

In the same manner, if we combine the collections A and C together we obtain the collection:

We can describe this symbolically as 3

This expression can't be symbolically abbreviated any further (if we are identifying the coins by denominations). In both situations, we still physically have seven distinct objects. Whether we can simplify the descriptions symbolically or not does not alter these physical facts.

Conclusion

Naming with symbols often provides new information and insight into vague and confusing situations—giving structure and game-changing clarity where there was none. Since this is similar to what physical light can do when it is showered upon a darkened landscape, we have metaphorically referred to this ability in symbols as a type of illumination.

Number naming like language naming is deeply interwoven into the fabric of our society, and on most days we usually encounter numerous situations involving its use (note that even basic things such as simply changing TV stations, watching the speedometer in our vehicle, or setting the heating time on a microwave oven have components that use number names).

We now conclude by recapping explicitly some of the benefits obtained through naming with numbers:

1. *Depth of expression*: Instead of seeing only crude notions such as more, longer, larger, or heavier in situations that are infinitely varied, numbers allow us to give names in depth, to all of that variety. Moreover, the naming is systematic, understandable, and maneuverable.

2. *Transformation of ideas*: Perhaps the most potent aspect of naming with numbers is the ability to systematically and unambiguously transform ideas. These transformations allow us to connect information in obvious as well new and emergent ways not always easy to see at the start.
   - This can be likened to the grammar in a language which gives us the ability to turn simple words (which are in and of themselves extraordinarily powerful) into a fuel that allows for the expression of the full range of life. Calculation in a sense allows us to almost symbolically liquefy phenomena—for ideas in symbolic and numerical form can, like fluids, naturally travel very far from their original sources while still retaining a powerful punch.

3. *Reproducibility of result*: Calculations and numerical representations have a well-defined, reproducible quality about them. Regardless of the time and the place or the symbols and methods employed, multiplication of the two natural numbers 52 and 86, as we have defined it, will always yield the value 4,472.
   - If a package of 500 ordered but un-numbered pages is dispersed by a gust of wind, the recipient will have a difficult enough task just knowing if he has retrieved them all much more how to put the papers back into the correct order. However, if the pages are numbered, he can systematically reproduce the original ordering of the pages and as a consequence know whether or not he has retrieved all of them. All of this from simply labeling the pages with numbers. Philip Davis and Reuben Hersh actually define mathematics in terms of reproducibility by saying: "The study of mental objects with reproducible properties is called mathematics."[10]

4. *Predictive power*: Naming with numbers gives us the ability to make predictions both through calculations and through direct representation.
   - *Calculation*: Calculation gives us the ability to make predictions on the cost of a trip, the number of chairs in an auditorium, the amount of money we earn in a year given an hourly wage, the distance around the earth, the equal allotment of portions among a group of people, and so on and so on.

- *Representation*: Language representation itself gives predictive power, for if someone told us that Redwoods National Park has exceptionally large and tall trees—we visit it with expectations. We would be incredulous and felt lied to if upon arrival we found a treeless barren desert instead.
- We have similar expectations if we were to hear that the temperature outside is 120°F, and would be shocked, if upon stepping outside, found that it was snowing. We continually make decisions on what to wear based on simple numbers that give the estimated temperatures for a given day. This is the predictive power of representation.
- Representation of quantity is such a valuable thing on its own that if that was all that numerals were useful for they would still be worth their weight in gold.
5. *Convenience of use with symbols*: It is much easier to manipulate and handle symbols than it is to directly handle the things they represent. This gives a convenience of use which is shocking in its reach. From the combination of reasoning and the simple handling of symbols on a piece of paper in front of us, we acquire the capacity to answer questions of monumental importance—questions that would often simply be impossible to answer without.

All of this and more from just the elementary aspects of numeration alone! There are so many more stories yet to tell, still even in arithmetic, let alone the rest of mathematics, but the twilight of this text is upon us so these tales will have to wait for another day.

13

General Résumé

The human brain runs on stories. Our theory of the world is largely in the form of stories. Stories are far more easily remembered and recalled than sequences of unrelated facts. . . . Thinking thrives on stories, on the construction and exploration of patterns of events and ideas.

—Frank Smith, contemporary psycholinguist, expert on
reading instruction and Albert Michotte, Belgian experimental
psychologist, author of *The Perception of Causality*[1]

Most men . . . speak primarily to narrate . . . to exhibit what they have undergone or seen. Cut us off from narrative, how would the stream of conversation, even among the wisest, languish into detached handfuls. . . . Strictly considered . . . knowledge is but recorded experience, and a product of history.

—Thomas Carlyle, nineteenth-century
Scottish historian, critic, and author[2]

IN CONCLUDING THE opening address to the Third Mediterranean Conference on Mathematics Education (2003), Apostolos Doxiadis eloquently stated:

Embed mathematics in the soul by embedding it in history, by embedding it in story. By showing how it is lovely and adventurous—the stuff of the best quest myths. By showing how it was created by complex, adventurous, brave, struggling human beings. . . . Give the poetry, the adventure and the problems, through stories, both small stories of environment and large stories of culture. Grip the heart—and the brain will follow.[3]

I agree. Of all of the techniques used in *How Math Works*, it is the telling of stories and the placing of situations in context which have served as the lubricant to try to make everything fit together. Throughout, I have tried to use the power of narrative and setting to walk through elementary arithmetic in ways that illuminate its wonder and power—in ways that, with any luck, breathe life back into this old and most familiar of subjects. Hopefully this tack has allowed readers to gain a greater appreciation of the fact that this gateway subject to all of mathematics is truly spectacular in its own right, and that in its everyday use we are in possession of ancient knowledge every bit as useful to us as the wheel, music, or even language itself.

Mathematics is a multifaceted, kaleidoscopic subject with a multitude of ways to view, define, and explain it. Couple this with the reality that there is much that we still don't know in regards to the true nature of mathematics, and one must constantly be on guard to remember that numerous paths and approaches (way too many to describe in several books much less a single one) must be used to describe it in all its subtleties. No one plan of attack can come close to fully encompassing this immense subject.

A path that I have taken is to view mathematics, at its most elementary, as a type of communication possessing all of the generality of traditional language—with numbers and calculations helping to describe the infinite range of quantity in its many guises in analogue to the way words and sentences help describe the diversity of everyday life.[4]

One that rather than sitting in isolation from the other ways in which we exchange and use information works in powerful conjunction with them. Taking this viewpoint is in no way demeaning to mathematics but suggests that it, like speech and writing themselves, is one of the highest endeavors that we collectively partake in—forming one of the major stories of mankind.

Symbolic systems were created in large measure to help us better navigate and control our own destiny in an extremely large and complicated world. Their great power is indebted, in part, to the advantages in maneuverability they offer over the less "malleable" and less "accessible" things they are called upon to represent. And one of the most potent demonstrations of this power continues to be through mathematics itself.

However, while sharing much in common with the traditional methods of communication and working in creative combination with them, mathematics is still not exactly the same thing as they—no more than language writing is the same thing as speech. And there is no denying that, since mathematics in general takes more effort to study than language, there are also huge psychological components tied in with its learning ("our ideas of education not mattering a single whit if we neglect to remember that we are in fact teaching human beings").[5]

Alfred North Whitehead stated in 1911 that, "Civilization advances by extending the number of important operations which we can perform without thinking about them."[6] The HA numerals stand out among written mathematical systems in that they allow large numbers of people (who are systematically taught them) to actually do this to a degree—acquiring a mastery, at least in whole number pursuits, that rivals the mastery acquired in language pursuits.

Oh, how we would like to see this in the learning of fractions, negative numbers, algebra, geometry, calculus, and mathematical reasoning in general. But once the symbols and rules from the various areas proliferate, interact, and no longer directly map to physical objects easily visualized, we see a majority of people hitting, in varying degrees, barriers to the effective understanding of higher-level concepts.

A goal of this book has been to tear down in some small part these barriers to understanding by attempting to shatter the "divinity of arithmetic," through showing that even the methods, which we now take most for granted, were not given to us from on high, but were actually the result of centuries of scientific efforts on the part of our predecessors.

My wish is that a conceptual currency in which to think about mathematics in general will be enhanced. And just as monetary currency is transportable across a broad swath of our modern-day economies, it is hoped that the conceptual currency aimed for here can be transported from these discussions on arithmetic into enhanced understanding in other areas of mathematics as well as in private demonstrations of arithmetical concepts to those we might tutor or instruct.

So, now as we conclude this tour, let's directly ask ourselves again: Why is this stuff important? Opinions are all over the map on this one, but I must confess to being appreciative of the spirit exhibited by John Perry, when he stated (at the 1901 British Association for the Advancement of Science Conference, in Glasgow, Scotland) that:

> I hold that the study (of mathematics) began because it was useful, it continues because it is useful, and it is valuable to the world because of the usefulness of its results. The pure mathematician must allow me to go on thinking that, if his discoveries were not being utilized continually, his study would long ago have degenerated into something like what the Aristotelian dialectic became in the fourteenth century.[7]

The beauty of mathematics is multilayered. For starters, there is the beauty of studying patterns in the math itself to come up with new methods and insights that allow us to see ever more deeply into mathematics—regardless of whether or not an application outside of the subject can be found. There is the beauty that many of these new methods and insights so attained may then allow us to accomplish tasks in a much more convenient fashion (e.g., methods for multiplying with small diagrams versus brute force repeated addition). There is the beauty that these new methods not only may make things more convenient for us to do but in many cases may make the decisive difference in whether or not we can accomplish the task at all.

There is yet another beauty that many of these new insights expressed with visible marks on paper are so much more than their scripted appearance. The symbols provide us not only with the means to peer deeper into the subject of mathematics itself but they also allow us to better describe and act upon the things that we need in our everyday life as well. And if that isn't enough, they additionally give us the extraordinary capability to discover monumental things about our past, our present, our future yea even the very universe itself! Many believe this to be the most astonishing thing of all.

Imagine, if you will, a beautiful painting that inspires awe in those who view it, simply for the masterpiece it is, yet at the same time also has something incredibly insightful to say about the inner workings of the sun as well as a million other apparently unrelated but significant things, and you catch a glimpse of why many rant and rave about the amazing nature of mathematics.

To some (perhaps the engineer and the mechanic), the magnificence of the automobile is in the beautiful mechanisms under the hood. To most, however, the beauty of the automobile is probably in the external appearance, the frills, and in the driving. Whatever your preference, you must admit that the splendor of the mechanisms would be much diminished, perhaps even marginalized, if that darn machine didn't actually rev up, move, and take you places.

Isn't it a far more encompassing beauty that "the wonderful things going on under the hood" gift us with the ability to do magnitudes more than simply admire at the magnificence of their inner workings? Isn't the mechanism far more enhanced by also gifting us with the amazing ability to dramatically impact ourselves and society to absolutely stupendous degrees? Dare I say that possibly the most beautiful thing of all about the automobile (as well as other types of motor vehicles) is the spectacular fact that the internal beauty of their workings opens wide to us the external beauty of our planet as well as our own potential to affect massive change?

Simply put, with motorized transport we are able to rapidly haul massive amounts of goods over long distances (bringing the wealth of the world to our doorsteps); quickly reach those in dire need of emergency assistance (saving their lives in many cases); and swiftly annihilate distances that would have taken our ancestors months if not forever to cover. Moreover, such transport allows us (from the beginnings of our humble driveways) to travel the length and breadth of entire continents exploring their wonders—doing so for nothing more than the sheer enjoyment of it all.

All of this in ways that owe greatly to the multitude who came before us. All of this in ways that would make them proud of what we have collectively accomplished as human beings.

I say ditto for mathematics!

Appendix A

THIS IS THE FIRST PAGE of examples that a student, who knew nothing of addition in HA numerals, would see or hear discussed by the teacher in explaining addition. No worked out examples on elementary addition were given at all. (*Schoolmaster's Assistant*, by Thomas Dilworth revised by R. Tagart, 1818 edition, 11.)

The Schoolmaster's Assistant. 11

L.	Yds.	Gals.	Tons.	Hhds.	lb.
4	43	764	3746	47476	461743
7	27	347	7436	73712	761780
3	39	387	3406	31819	476332
2	13	736	7398	41243	126722
3	37	397	3373	71208	310748
7	46	473	4731	70956	571388
6	23	382	2264	81465	704714
4	59	769	4731	31269	312624
7	94	367	7169	74196	781462

Miles.	Leagues.	Years.
4734736	46431734	347312484
3474312	71261374	168126312
2546325	92652724	718125191
7369138	86337266	731618191
3143618	74147312	312134716
4733216	47312614	873263298
2473347	27467573	312614712
3712612	31216126	977647829
5726384	39874129	312814796

OF COMPOUND ADDITION.

Q. WHAT is compound Addition ?

A. Compound Addition is the adding of several numbers together, having divers denominations.

1. OF MONEY.

Q. Which are the denominations of English money ?

A. 4 Farthings make 1 Penny.

12 Pence ——— 1 Shilling.

20 Shillings ——— 1 Pound sterling.

Q. Are there no other names of money used in England ?

A. Yes : such as,

		l.	s.	d.
A Moidoire	=	1	7	0
A Guinea	=	1	1	0
A Half Guinea	=	0	10	6
A Crown	=	0	5	0
A Half crown	=	0	2	6

There are also several smaller pieces which speak their own value, as a six-pence, four-pence, three-pence, two-pence, penny, half-penny, and farthing.

B 2

Appendix B

L ETTER FROM GEORGE WASHINGTON TO NICHOLAS PIKE, June 20, 1788.

MOUNT VERNON, June 20, 1788.

SIR: I request you will accept my best thanks for your polite letter of the 1st of January (which did not get to my hand 'till yesterday) — and also for the copy of your *System of Arithmetic* which you were pleased to present to me.

The handsome manner in which that work is printed, and the elegant manner in which it is bound are pleasing proofs of the progress which the arts are making in this Country. But I should do violence to my own feelings, if I suppressed an acknowledgement of the belief that the work itself is calculated to be equally useful and honorable to the United States.

It is but right, however, to apprise you, that, diffident of my own decision, the favourable opinion I entertain of your performance is founded rather on the explicit and ample testimonies of Gentlemen confessedly possessed of great mathematical knowledge, than on the partial and incompetent attention I have been able to pay to it myself. But I must be permitted to remark that the subject, in my estimation, holds a higher rank in the literary scale than you are disposed to allow. The science of figures, to a certain degree, is not only indispensably requisite in every walk of civilised life, but the investigation of mathematical truths accustoms the mind to method and correctness in reasoning, and is an employm. peculiarly worthy of rational beings. In a cloudy state of existence, where so many things appear precarious to the bewildered research, it is here that the rational faculties find a firm foundation to rest upon. From the high ground of Mathematical & Philosophical demonstration, we are insensibly led to far nobler speculations & sublime meditations.

I hope and trust that the work will ultimately prove not less profitable than reputable to yourself. It seems to have been conceded, on all hands, that such a system was much wanted. Its merits being established by the approbation of competent judges, I flatter myself that the idea of its being an American production, and the first of the kind which has appeared; will induce every patriotic and liberal character to give it all the countenance & patronage in his power. In all events, you may rest assured, that, as no person takes more interest in the encouragement of American genius, so no one will be more highly gratified with the success of your ingenious, arduous & useful undertaking than he, who has the unfeigned pleasure to subscribe himself with esteem & regard

Sir Your Most Obed' and Very H^{ble} Servant,

G:° WASHINGTON

NICHOLAS PIKE, Esq'

Appendix C

The Fifteen Most Significant Swiss of All-Time

(According to a 2008 Survey conducted by Swiss newspaper *SonntagsZeitung*)

1. Albert Einstein (1879–1955), physicist
2. Gottlieb Duttweiler (1888–1962), businessman and philanthropist
3. Roger Federer (1981–), tennis player
4. Johann Heinrich Pestalozzi (1746–1827), educationalist
5. Henry Dunant (1828–1910), founder of the Red Cross
6. Paracelsus (1493–1541), physician
7. Nicolas Hayek (1928–), businessman
8. Claude Nicollier (1944–), astronaut
9. Alfred Escher (1819–1882), entrepreneur, railway pioneer
9. Leonhard Euler (1707–1783), mathematician
11. Friedrich Dürrenmatt (1921–1990), writer and artist
11. Le Corbusier (1887–1965), architect
13. Jean-Jacques Rousseau (1712–1778), philosopher and writer
14. Mani Matter (1936–1972), singer
15. Henri Guisan (1874–1960), general

Appendix D

F*IRST LESSONS IN ARITHMETIC ON THE PLAN OF PESTALOZZI*, 2nd ed. (Boston: Cummings and Hilliard, 1822), 1, 3.

PART I.

SECTION I.

A* 1. How many thumbs have you on your right hand? how many on your left? how many on both together?

2. How many hands have you ?

3. If you have two nuts in one hand and one in the other, how many have you in both ?

4. How many fingers have you on one hand ?

5. If you count the thumb with the fingers, how many will it make ?

6. If you shut your thumb and one finger and leave the rest open, how many will be open ?

7. If you have two cents in one hand, and two in the other, how many have you in both ?

Sect. I.   ARITHMETIC.   3

24. Five and four are how many ?

25. How many fingers have you on both hands ?

26. Four and four are how many ?

27. How many fingers and thumbs have you on both hands ?

28. Five and five are how many ?

29. If you had six marbles in one hand, and four in the other ; how many would you have in the one, more than in the other ? how many would you have in both hands ?

30. Six and four are how many ?

31. David had seven nuts, and gave three of them to George, how many had he left?

32. Seven less three are how many?

33. Two boys, James and Robert, played at marbles; when they began, they had seven apiece, and when they had done, James had won four; how many had each then?

34. Seven and four are how many?

35. Seven less four are how many?

36. A boy, having eleven nuts, gave away three of them, how many had he left?

37. Eleven less three are how many?

38. If you had eight cents, and your papa should give you five more, how many would you have?

39. Eight and five are how many?

40. A man bought a sheep for eight dollars, and a calf for seven dollars, what did he give for both?

Notes

Introduction

1. Ian Stewart, *From Here to Infinity* (New York: Oxford University Press, 1996), 1.

2. Hardy Grant, "A Sketch of the Cultural Career of Mathematics," in *Essays in Humanistic Mathematics*, ed. Alvin White (Washington, D.C.: Mathematical Association of America, 1993), 23.

3. Amalia E. Gnanadesikan, *The Writing Revolution: Cuneiform to the Internet* (Oxford: Blackwell-Wiley, 2008), 4–6.

4. George Armitage Miller, *The Science of Words* (New York: Scientific American Library, 1991), 2.

5. Steven Pinker, *The Language Instinct: How the Mind Creates Language* (New York: Perennial Classics, 2000), 155.

6. Colleen Donnelly, *Linguistics for Writers* (Albany: State University Press of New York, 1994), 3.

7. Lancelot Hogben, *Mathematics for the Million* (New York: W. W. Norton, 1993), 14.

Part I

1. Benjamin Farrington, *Greek Science: Its Meaning for Us* (Baltimore: Penguin Books, 1969), 311.

2. John Fauvel and Jan Van Maanen, eds. *History in Mathematics Education: The ICMI Study* (Dordrecht: Kluwer Academic Publishers, 2000), 36.

Chapter 1

1. Francis Bacon, *The New Organon*, eds. Lisa Jardine and Michael Silverthorne (Cambridge: Cambridge University Press, 2000), 33.

2. James Ward, "Notes on Education Values," *Journal of Education, A Monthly Record and Review 22* (January–December 1890): 630.

3. Karen Wynn, "Addition and Subtraction by Human Infants," *Nature 358* (1992): 749–50.

4. Ann Wakeley, Susan Rivera, and Jonas Langer, "Not Proved: Reply to Wynn," *Child Development* 71 (2000): 1537–39.

5. Note that we have replaced the | tally mark by the symbol (|). This change is purely cosmetic and is done for the sake of consistency.

6. Florian Cajori, *A History of Mathematical Notations: Two Volumes Bound as One* (New York: Dover, 2007), 31–34.

Chapter 2

1. Guy Deutscher, *The Unfolding of Language: An Evolutionary Tour of Man's Greatest Invention* (New York: Henry Holt, 2005), 1.

2. David R. Olson, *The World on Paper: The Conceptual and Cognitive Implications of Writing and Reading* (New York: Cambridge University Press, 1996), xv.

3. Deutscher, *The Unfolding of Language*, 5–15.

4. Kyra Karmiloff and Annette Karmiloff-Smith, *Pathways to Language: From Fetus to Adolescent* (Cambridge, Mass.: Harvard University Press, 2001), 4–5.

5. Denise Schmandt-Besserat, *How Writing Came About* (Austin: University of Texas Press, 1996), 98. Schmandt-Besserat further suggests that the creation of numerals preceded by several millennia language writing and was in part responsible for its invention. Mapping the progression onto our discussion, there was evidently a long period where token numerals were used (this would roughly match our use of rocks to measure collections) and then came the purely written and abstract numerals (these would roughly align with our coin numerals). This theory has endured criticism; Geoffrey Sampson, *Writing Systems: A Linguistic Introduction* (Stanford, Calif.: Stanford University Press, 1985), 57–61.

6. George Yule, *The Study of Language*, 2nd ed. (New York: Cambridge University Press, 1996), 9.

7. Sampson, *Writing Systems*, 30.

8. Barry B. Powell, *Writing: Theory and History of the Technology of Civilization* (Chichester, UK: Wiley-Blackwell, 2009), 11.

9. In discussing some of the advantages and disadvantages here, we have blended together what many linguists prefer to distinguish as two types of writing: glottographic writing and semasiographic writing. Glottographic writing is the type of writing that visibly captures much of spoken communication—what we have called language writing. Semasiographic writing corresponds to other types of writing used for communication (such as occurs in mathematics, with systems of road signs, or pictorial systems in general) but are not directly related to the spoken language of the users (Sampson, *Writing Systems*, 29–30). Such blending does no vandalism to our discussion here.

10. Victoria Fromkin, Robert Rodman, and Nina Hyams, *An Introduction to Language*, 9th ed. (Boston: Wadsworth, 2011), 543.

11. Roy Harris, *Rethinking Writing* (London: Continuum, 2001), xii–xiii.

12. Harris, *Rethinking Writing*, xiii.

13. Roy Harris, *Signs, Language, and Communication* (New York: Routledge, 1996), 184–85.

14. David R. Olson, *The World on Paper: The Conceptual and Cognitive Implications of Writing and Reading* (New York: Cambridge University Press, 1996), 164–66.

15. Olson, *The World on Paper*, 86.

16. Olson, *The World on Paper*, 111.

17. Marion de Lemos, *Closing the Gap between Research and Practice: Foundations for the Acquisition of Literacy* (Camberwell: Australian Council for Educational Research, 2002), 3.

18. The couching of this very fundamental notion in alphabetic terms is in no way a statement hailing the superiority of alphabetic writing systems over other systems of language writing. It is used here as well as throughout the rest of the book as an aid in supplying key conceptual insight to readers who are very familiar with alphabets.

19. In some cases, we use an existing word and simply give it an additional meaning. Such words that have multiple and distinctive meanings and yet are spelled the same are called homographs and include the following: lead, bear, punch, crop, and plane.

20. Henri Lebesgue, *Measure and the Integral*, ed. Kenneth O. May (San Francisco: Holden-Day, 1966), 18.

Chapter 3

1. Howard W. Eves, *Mathematical Circles Squared* (Boston: Prindle, Weber and Schmidt, 1972), 13.

2. This situation can happen in language as well. In certain parts of the world, there are people who use the same writing system yet don't speak the same language. Also throughout the Middle Ages and into the Renaissance, European scholars of varying mother tongues chose to communicate by writing in Latin.

3. Stephen Chrisomalis, "The Cognitive and Cultural Foundations of Numbers," in *The Oxford Handbook of the History of Mathematics*, eds. Eleanor Robson and Jacqueline Stedall (Oxford: Oxford University Press, 2009), 506; Georges Ifrah's book, *The Universal History of Numbers: From Prehistory to the Invention of the Computer* (New York: Wiley, 2000), gives a wealth of examples of different numeral systems.

4. John W. Wright, ed. *The New York Times Guide to Essential Knowledge* (New York: St. Martin's Press, 2011), 394.

5. Lam Lay Yong and Ang Tian Se, *Fleeting Footsteps: Tracing the Conception of Arithmetic and Algebra in Ancient China*, rev. ed. (Singapore: World Scientific Publishing, 2004), 47–48.

6. There has been some debate over the years as to whether or not some of the smaller digits contain pictorial cues to their value.

7. Christopher B. Steiner, "Authenticity, Repetition, and the Aesthetics of Seriality: The Work of Tourist Art in the Age of Mechanical Reproduction," in *Unpacking Culture: Art and Commodity in Colonial and Postcolonial Worlds*, edited by Ruth B. Phillips and Christopher B. Steiner (Berkeley: University of California Press, 1999), 100–101.

8. A fine of 20 soldi was assessed to those who broke the law in Florence. The concerns are explained in an old Venetian work on bookkeeping, "the old figures alone are used because they cannot be falsified as easily as those of the new art of computation, of which one can with ease make one out of another, such as turning the zero into a 6 or a 9, and similarly many others can also be falsified." Karl Menninger, *Number Words and Number Symbols: A Cultural History of Numbers* (Cambridge, Mass.: MIT Press, 1969), 426–27.

9. Ifrah, *The Universal History of Numbers*, 341.

10. J. J. O'Connor and E. F. Robertson, "Brahmagupta," on *MacTutor History of Mathematics*, 2000, www-history.mcs.st-andrews.ac.uk/Biographies/Brahmagupta.html.

11. Richard Feynman, *Six Easy Pieces* (New York: Basic Books, 2011), 69.

Chapter 4

1. Alfred North Whitehead, *An Introduction to Mathematics* (New York: Henry Holt, 1911), 11.

2. For the value of Pi, see Petr Beckmann, *A History of Pi* (New York: St. Martin's Press, 1971), 22. As discussed earlier, it would not be until the early centuries CE before the precursors to the symbols {1, 3} would enter into use as whole numbers and not until the late 1500s of the current era before they would be employed as decimal fractions. Using them here, however, causes no loss of generality and they have been included for the sake of clarity for the reader.

3. ≈ means "approximately equal to."

4. Nigel Wilson, ed. *Encyclopedia of Ancient Greece* (New York: Routledge, 2006), 269.

5. A great circle on a sphere is a circle whose center is also the center of the sphere.

6. Syene is not located on the equator; nevertheless our analysis holds due to the fact that the angle between the sun's rays and the tower is 0.

7. Eratosthenes used a unit called a stadia rather than miles to measure distances and there is some uncertainty as to the true length of a stadia. According to some values given, Eratosthenes's number is remarkably accurate and according to others his measurement was off a bit. In any event, the major difficulty lies with whether the distance from Alexandria to Syene was accurate, not the method used, which is entirely sound for a spherical earth.

8. Raymond G. Ayoub, ed. *Musings of the Masters: An Anthology of Mathematical Reflection* (Washington, D.C.: Mathematical Association of America, 2004), 151.

9. Jacob Klein, *Greek Mathematical Thought and the Origin of Algebra* (Cambridge, Mass.: MIT Press, 1968), 208.

10. John de Pillis, *777 Mathematical Conversation Starters* (Washington, D.C.: Mathematical Association of America, 2002), 197.

Part II

1. Alfred W. Crosby, *The Measure of Reality: Quantification and Western Society, 1250–1600* (New York: Cambridge University Press, 1997), 113.

Author's Note

1. Quote attributed to Wolfgang Ratke (1571–1635), in Robert Hebert Quick, *Essays on Educational Reformers* (New York: D. Appleton, 1897), 114.

Chapter 5

1. Sir Michael Atiyah, "Special Article Mathematics in the 20th Century," *Bulletin of the London Mathematical Society* 34 (2002): 1–15.

2. Abaci come in a wide variety of different models some involving rods and beads and some using other arrangements (such as columns, tokens, and chips). It is not clear to historians the ancestral type of abacus or calculating device used in ancient India to create the HA numerals.

3. If the commutativity of addition is taken into account (e.g., $1 + 2 = 2 + 1$ or $8 + 9 = 9 + 8$), these one hundred possibilities can be shortened to fifty-five.

4. It is possible to subtract a larger whole number from a smaller whole number but the result will be a negative number as opposed to a whole number. Negative numbers are valid entities in their own right and turn out to be very useful throughout mathematics. They will not be investigated in this book, however.

5. Susan Ross and Mary Pratt-Cotter, "Subtraction in the United States: An Historical Perspective," *The Mathematics Educator* 8 (1997): 4–8.

Chapter 6

1. Gilbert Keith Chesterson, *Tremendous Trifles* (New York: Dodd, Mead & Co., 1920), 245–46.

2. Voicing out larger numbers such as 1,235,372 will take longer than a second on average.

3. Here we literally mean 2 – 6 with no tens to borrow from. To calculate this involves the use of signed whole numbers (integers) which are not discussed in this book.

4. Note that the sum of the numbers in the top row equals 2047.

5. James Stuart Tanton, *Encyclopedia of Mathematics* (New York: Facts on File, 2005), 446.

6. A variant of the Egyptian method of multiplication was used in Russia for centuries. It is called Russian peasant multiplication.

7. Since "m" can take on values that are not whole numbers (in the equation $E = mc^2$), the multiplication in this equation has a more general interpretation than the one given in this chapter.

Chapter 7

1. Friedrich Wilhelm Nietzsche, *Twilight of the Idols: with the Antichrist and Ecce Homo*, trans. Antony M. Ludovici (Hertfordshire, UK: Wordsworth Editions Limited, 2007), 47.

2. Many other nations multiply using different methods from the ones presented here.

3. Ernesto Estrada, *The Structure of Complex Networks: Theory and Applications* (New York: Oxford University Press, 2011), 3.

4. Estrada, *The Structure of Complex Networks*, 3–4.

Chapter 8

1. Alfred North Whitehead, *An Introduction to Mathematics* (New York: Henry Holt, 1911), 59.

2. Such twin situations involving an interchange of values where first the 3 is on the groups and the 8 is on the objects and then the 8 is on the groups and the 3 is on the objects are sometimes called reciprocal relationships. These relationships occur in much deeper ways in certain branches of mathematics and physics such as fractions, number theory, group theory, and electromagnetism.

3. Florian Cajori, *A History of Mathematical Notations: Two Volumes Bound as One* (New York: Dover, 2007), 268–70.

4. Cajori, *A History of Mathematical Notations*, 270.

5. "Merriam-Webster Online Dictionary," 2012, www.merriam-webster.com.

6. Cajori, *A History of Mathematical Notations*, 271–72.

Chapter 9

1. Steven Leinwand, "It's Time to Abandon Computational Algorithms," *Education Week*, 13 (1994), 36.

2. David Klein and R. James Milgram, *The Role of Long Division in the K–12 Curriculum*, February 2000, www.csun.edu/~vcmth00m/longdivision.pdf.

3. David Eugene Smith, *The History of Mathematics*, vol. 2, ed. Eva May Luse Smith (New York: Dover, 1958), 140–43.

4. "Britannica Online Encyclopedia," 2012, www.britannica.com.

5. This represents the galley division 965347653446 ÷ 6543218 = 147534. Remainder 529034 (not all digits are scratched out here). Frank Swetz and Victor Katz, "Opus Arithmetica of Honoratus," in *Mathematical Treasures* (Mathematical Sciences Digital Library: Mathematical Association of America, 2012).

6. Frank Swetz, *Capitalism and Arithmetic: The New Math of the 15th Century* (LaSalle, Ill.: Open Court, 1987), 214.

7. Lam Lay Yong, "On the Chinese Origin of the Galley Method of Arithmetical Division," *British Journal of the History of Science* 3 (1966): 66–69.

8. Lambert Lincoln Jackson, *The Educational Significance of 16th Century Arithmetic: From the Point of View of the Present Time* (New York: Teacher's College, Columbia University, 1906), 69.

9. Harald Ness, "Mathematics: An Integral Part of Our Culture," in *Essays in Humanistic Mathematics*, ed. Alvin White (Washington, D.C.: Mathematical Association of America, 1993), 49.

Part III

1. Robert Zemeckis, *Contact* (Warner Brothers, 1997).

2. Carl C. Gaither and Alma E. Cavazos-Gaither, eds. *Gaither's Dictionary of Scientific Quotations Second Edition* (New York: Springer, 2012), 1335.

Chapter 10

1. John Amos Comenius, *The Great Didactic of John Amos Comenius*, M. W. Keatinge trans. (London: Adam and Charles Black, 1896), 218.

2. Robert Hebert Quick, *Essays on Educational Reformers* (New York: D. Appleton, 1897), 82.

3. "Math Made Interesting," *Time*, September 22, 1961, 59.

4. This statement is speaking to the general attitudes of not an insignificant number of research mathematicians toward teaching which, in its rawest form, is perhaps best encapsulated by the words of British mathematician G. H. Hardy who in 1940 wrote that, "Exposition, criticism, appreciation, is work for second-rate minds." G. H. Hardy, *A Mathematician's Apology* (Cambridge: Cambridge University Press, 1999), 61. It is not meant to imply that there were not then nor are not now research mathematicians who do care a great deal about K–12 education, as well as freshman and sophomore college education. In fact, today we see many research mathematicians, including several members of the National Academy of Sciences, who are very seriously involved in mathematics education. They should be applauded for this.

5. Henry O. Pollack, "A History of Teaching Modeling," in *A History of School Mathematics*, George M. A. Stanic and Jeremy Kilpatrick, eds. (Reston, Va.: National Council of Teachers of Mathematics, 2003), 665.

6. Takashi Kojima, *The Japanese Abacus: Its Use and Theory* (Tokyo: Tuttle, 1954), 4–5.

7. Bead abaci such as the ones discussed in this book both predate and postdate medieval counter abaci. The Roman hand abacus is a bead-like abacus dating back to the days of the empire and modern abaci such as the Soroban and Suan Pan are of the bead variety. A device similar to the counter abacus was discovered on the Greek island of Salamis in the nineteenth century. It is believed to date back at least to the third century BCE.

8. Karl Menninger, *Number Words and Number Symbols: A Cultural History of Numbers*, (Cambridge, Mass.: MIT Press, 1969), 327; Florian Cajori, *A History of Mathematics*, 2nd ed. (New York: Macmillan, 1919), 116–21.

9. Frank Swetz, *Capitalism and Arithmetic: The New Math of the 15th Century* (LaSalle, Ill.: Open Court, 1987), 188.

10. Swetz, *Capitalism and Arithmetic*, 29.

11. A third type of reckoning used at this time was called finger reckoning. It was an elaborate system of representing numbers by various finger and hand symbols and was often employed to perform simple calculations, especially in the marketplace. The use of this type of reckoning, which dates way back, was very important, and was widely discussed in books of the time. Here, however, it is ancillary to our treatment and will not be discussed.

12. Ari Ben-Menahem, *Historical Encyclopedia of Natural and Mathematical Sciences*, Vol. 1 (Berlin: Springer-Verlag, 2009), 5731.

13. Reportedly, the French mathematician, Paul Tannery, after much practice with an ancient Greek system of numerals, became quite proficient at multiplication using them. Vera Sanford, *A Short History of Mathematics*, ed. John Wesley Young (Boston: Houghton Mifflin, 1930), 87.

14. J. J. O'Connor and E. F. Robertson, "Arabic Mathematics: Forgotten Brilliance?" on *Mac-Tutor History of Mathematics*, 1999, www-history.mcs.st- and.ac.uk/HistTopics/Arabic_mathematics.html.

15. Peter Munoz, *Frederick Barbarossa: A Study in Medieval Politics* (Ithaca, N.Y.: Cornell University Press, 1969), 310–12.

16. Laurence Sigler, *Fibonacci's Liber Abaci: A Translation into Modern English of Leonardo Pisano's Book of Calculation* (New York: Springer, 2002), 129, 156, 213, 228.

17. Swetz, *Capitalism and Arithmetic*, 12.

18. Paul F. Grendler, *Schooling in Renaissance Italy, Literacy and Learning 1300–1600* (Baltimore: Johns Hopkins University Press, 1991), 308.

19. Ronald G. Witt, *In the Footsteps of the Ancients: The Origins of Humanism from Lovato to Bruni* (Leiden: Brill, 2000), 194–95.

20. Paul F. Grendler, *The Universities of the Italian Renaissance* (Baltimore: Johns Hopkins University Press, 2002), 426; Grendler, *Schooling in Renaissance Italy*, 309; William Eamon, "Markets, Piazzas, and Villages," in *The Cambridge History of Science: Early Modern Science*, vol. 3, Katherine Park and Lorraine Daston, eds. (Cambridge: Cambridge University Press, 2006), 212.

21. Timothy J. Reiss, *Knowledge, Discovery and Imagination in Early Modern Europe: The Rise of Aesthetic Rationalism* (Cambridge: Cambridge University Press, 1997), 140.

22. Grendler, *Schooling in Renaissance Italy*, 316.

23. Counter reckoning was still carefully described, alongside computation with the HA numerals in pen, in the German and English arithmetic books well into the sixteenth century. Sanford, *A Short History of Mathematics*, 93.

24. The difficulties in accepting zero as a legitimate entity were not trivial, and offer a vivid illustration of humankind's long conflict over what a symbol really is. Is it merely a representative of the object/idea or is it inextricably paired, part and parcel, with what it stands for? If it is the latter, then the idea of "nothing" means just that, nothing, and a symbol of any kind is actually "something" and their inextricable pairing means that "nothing" is paired with "something."

This represented a conflict for many exposed to the idea of zero throughout the Middle Ages. While this conflict may seem trivial to us today, we still have our own battles to fight concerning the relationships between symbols and reality. We continue to wrestle with what the symbols and processes of modern physics really tell us about the physical world. Do they just allow us to make predictions or do they provide real descriptions of the external physical world? Unlike the medievalists however and with the benefit of history on our side, we more willingly push forward, sometimes, with awe-inspiring results.

25. Grendler, *Schooling in Renaissance Italy*, 307.

26. Tryon Edwards, *A Dictionary of Thoughts: Being a Cyclopedia of Laconic Quotations from the Best Authors of the World, Both Ancient and Modern* (Detroit: F. B. Dickerson, 1908), 441.

27. Henry Holman, *Pestalozzi: An Account of His Life and Work* (London: Longmans, Green, 1908), 4.

28. Charles Issawi, "Europe, the Middle East and the Shift in Power: Reflections on a Theme by Marshall Hodgson," *Comparative Studies in Society and History 22* (1980), 492; Brian Moynahan, *God's Bestseller: William Tyndale, Thomas More, and the Writing of the English Bible—A Story of Martyrdom and Betrayal* (New York: St. Martin's Press, 2003), 69.

29. Lucien Febvre and Henri-Jean Martin, *The Coming of the Book: The Impact of Printing 1450–1800*, trans. David Gerard (London: Verso, 1997), 186, 262.

30. Swetz, *Capitalism and Arithmetic*, 33; Augustus De Morgan, *Arithmetical Books from the Invention of Printing to the Present Time: Being Brief Notices of a Large Number of Works Drawn Up from Actual Inspection* (London: Taylor and Walton, 1847), v–vi.

31. David Eugene Smith, *A Source Book for Mathematics: 125 Classic Selections from the Writings of Pascal, Leibniz, Euler, Fermat, Gauss, Descartes, Newton, Riemann and Many Others* (New York: Dover, 1959), 1–2.

32. In terms of raw speed, the bead abacus is clearly a superior device to the counter abacus. While the counter abacus certainly offered speed advantages over calculating directly with Roman numerals in writing, I didn't find clear and convincing evidence that the counter abacus had such an advantage over calculating in writing with HA numerals. The small amount of evidence that I am aware of gives conflicting testimony on this issue. Conceivably the need to slide counters around as well as having to borrow them from the pile could have taken as much time as doing the same calculation with HA numerals—even in the hands of a master abacist. And since master counter abacists are at present an extremely rare if not extinct breed (as opposed to bead abacus wizards such as Mr. Matsuzaki), it will be difficult to truly gauge the speed of such abaci in a modern day competition.

33. John Denniss, "Learning Arithmetic: Textbooks and Their Users in 1500–1900," in *The Oxford Handbook of the History of Mathematics*, eds. Eleanor Robson and Jacqueline Stedall (Oxford: Oxford University Press, 2009), 451.

34. Lam Lay Yong and Ang Tian Se, *Fleeting Footsteps: Tracing the Conception of Arithmetic and Algebra in Ancient China*, Rev. ed. (Singapore: World Scientific Publishing, 2004), chap. 9.

35. Lam Lay Yong, "The Development of Hindu-Arabic and Traditional Chinese Arithmetic," *Chinese Science 13* (1996): 45.

36. Ron Howard, *Apollo 13* (Universal Pictures, 1995).

Chapter 11

1. Sir William Petty, "Plan of a Trade or Industrial School," *The American Journal of Education* 11 (1862), 199–200. (Originally Printed in 1647.)

2. Roger Bucher, Dale Griffin and Johanna Peetz, "The Planning Fallacy: Cognitive, Motivations, and Social Origins," in *Advances in Experimental Social Psychology 43*, eds. Mark P. Zanna and James M. Olson (Amsterdam: Elsevier, 2010), 19.

3. The discussion that follows is not meant to imply that there was anything even close to a centralized group of educators, on a national scale, who were on task and deliberately asking these questions at this time. Education in Europe was highly disorganized, often exclusive, localized, and hit or miss (Paul F. Grendler, *Books and Schools in the Italian Renaissance* [Hampshire: Variorum, 1995], 774–81). However, even if local educators were not consciously asking these questions, they still had to deal with them on some level and our intent is to show linkage between some of the issues facing educators during this period with some of the issues facing educators now or during any period.

4. Later reformers of education spoke out vigorously in their condemnation of this latter practice of "punishing for failing to understand quickly enough."

5. John Denniss, "Learning Arithmetic: Textbooks and Their Users in 1500–1900," in *The Oxford Handbook of the History of Mathematics*, eds. Eleanor Robson and Jacqueline Stedall (Oxford: Oxford University Press, 2009), 449.

6. Augustus de Morgan, *Arithmetical Books from the Invention of Printing to the Present Time: Being Brief Notices of a Large Number of Works Drawn Up from Actual Inspection* (London: Taylor and Walton, 1847), xxi.

7. Florian Cajori, *A History of Elementary Mathematics with Hints on Methods of Teaching* (New York: Macmillan, 1929), 210–11.

8. Lewis Flint Anderson, *History of Common School Education* (New York: Henry Holt, 1909), 200.

9. The issuing of these statutes also suggests that mathematics, including arithmetic in some form, had been part of English university curricula up to this time. Walter William Rouse Ball, *A History of the Study of Mathematics at Cambridge* (Cambridge: Cambridge University Press, 1889), 12–13. This is also supported by the existence of handwritten/pre-printing press works in old English such as *The Craft of Nombrynge* (ca. 1400s) and *The Art of Nombrynge* (ca. 1400s) based on earlier works from the thirteenth century, Robert Steele, *Earliest Arithmetics in English* (London: Oxford University Press, 1922), v–vi.

10. Geoffrey Howson, *A History of Mathematics Education in England* (Cambridge: Cambridge University Press, 1982), 35–38.

11. Howson, *A History of Mathematics Education in England*, 29, 41.

12. These statements are not meant to be disparaging in any way of Pepys. As Chief Secretary to the Admiralty, he had critical influence in transforming England into becoming the preeminent naval power in the world. It is to his determination, that upon realizing that his arithmetic knowledge was lacking, he decided to correct this, with the help of a tutor, in his late twenties—personally presaging the idea of adult continuing education by centuries. It is to his vision, that upon realizing that the entire Royal Navy was lacking in mathematical knowledge, he sought to correct it by establishing the Royal Mathematical School at Christ's Hospital.

13. George Emery Littlefield, *Early Schools and School-Books of New England* (Boston: Club of Odd Volumes, 1904), 167.

14. Benjamin Franklin, *Autobiography of Benjamin Franklin*, ed. Charles W. Eliot (New York: P. F. Collier, 1909), 14.

15. Littlefield, *Early Schools and School-Books of New England*, 173.

16. What we today call common fractions were referred to as vulgar fractions during this period. Thus, vulgar arithmetic is referring to the arithmetic of common fractions. Many of my twenty-first-century students tell me they find the older terminology more appropriate.

17. Nerida Ellerton and Ken Clements, "The Process of Decolonizing School Mathematics Textbooks and Curricula in the United States," *ICME* 11, TSG 38 (2008): 1.

18. Patricia Cline Cohen, "Numeracy in Nineteenth Century America," in *A History of School Mathematics*, eds. George M. A. Stanic and Jeremy Kilpatrick (Reston, Va.: National Council of Teachers of Mathematics, 2003), 52.

19. Extending this idea to include the entire citizenry would be slow in coming, and is, of course, one of the major stories in the history of this nation. It is still a work in progress to this very day.

20. Littlefield, *Early Schools and School-Books of New England*, 179–82.

21. Cohen, "Numeracy in Nineteenth Century America," 44.

22. Horace Mann, ed. *The Common School Journal* 5 (1843): 119.

23. Cohen, "Numeracy in Nineteenth Century America," 55.

24. Samuel Griswold Goodrich, *The Child's Arithmetic Being an Easy and Cheap Introduction to Daboll's, Pike's, White's and other Arithmetics* (Hartford: Samuel Griswold Goodrich, 1818), iii.

25. Robert Chambers, ed. *Cyclopaedia of English Literature: A History, Critical and Biographical of British Authors, From the Earliest to the Present Times.* Vol. 1 (Edinburgh: W. & R. Chambers, 1858), 256.

26. Robert Quick summarizes some of Comenius's thoughts from the *Didactica Magna* (*The Great Didactic*): "If we would ascertain how teaching and learning are to have good results, we must look to the known processes of Nature and Art. A man sows seed, and it comes up he knows not how, but in sowing it he must attend to the requirements of Nature. Let us then look to Nature to find out how knowledge takes root in young minds. We find that Nature waits for the fit time. Then too, she has prepared the material before she gives it form. In our teaching we constantly run counter to these principles of hers. We give instruction before the young minds are ready to receive it. We give the form before the material. Words are taught before the things to which they refer. When a foreign tongue is to be taught, we commonly give the form, i.e., the grammatical rules, before we give the material, i.e., the language, to which the rules apply. We should begin with an author, or properly prepared translation-book, and abstract rules should never come before examples." Quick, *Essays on Educational Reformers*, 136.

27. Will Seymour Monroe, *Comenius and the Beginnings of Educational Reform* (New York: Charles Scribner, 1900), 49, 53, 56–58.

28. James B. Conant, "Comenius and Harvard," in *Teacher of Nations: Addresses and Essays in Commemoration of the Visit to England of the Great Czech Educationalist Jan Amos Komenský, Comenius, 1641–1941*, ed. Joseph Needham (Cambridge: Cambridge University Press, 1942), 40.

29. Paul Monroe, ed. *A Cyclopedia of Education.* Vol. 5 (New York: Macmillan, 1913), 214.

30. Quick, *Essays on Educational Reformers*, 251.

31. Edward Craig, ed. *Routledge Encyclopedia of Philosophy.* Vol. 8 (London: Routledge, 1998), 372.

32. Henry Jones, "The Social Organism," in *The British Idealists*, ed. David Boucher (New York: Cambridge University Press, 1997), 8.

33. Tryon Edwards, *A Dictionary of Thoughts: Being a Cyclopedia of Laconic Quotations from the Best Authors of the World, Both Ancient and Modern* (Detroit: F. B. Dickerson, 1908), 149.

34. Albert Einstein, *Cosmic Religion with Other Opinions and Aphorisms* (New York: Covici-Friede, 1931), 97.

35. Hermann Krusi Jr., *Pestalozzi* (Carlisle, Mass.: Applewood Books, 1875), 16–22.

36. Roger de Guimps, *Pestalozzi: His Life and Work.* 2nd ed., trans. J. Russell (New York: D. Appleton, 1890), viii, 218, 313–14.

37. Guimps, *Pestalozzi*, 174.

38. Johann Heinrich Pestalozzi, *How Gertrude Teaches Her Children: An Attempt to Help Mothers to Teach Their Own Children and an Account of the Method—A Report to the Society of*

the Friends of Education, Burgdorf, 2nd ed., trans. Lucy E. Holland and Frances C. Turner, ed. Ebenezer Cook (Syracuse: C. W. Bardeen, 1898), 60–61.

39. Pestalozzi, *How Gertrude Teaches Her Children,* 126.

40. Pestalozzi, *How Gertrude Teaches Her Children,* 97.

41. There is some disagreement on exactly what Pestalozzi meant when he used the term *Anschauung.* Whatever his complete purposes in using the word, it certainly includes the coalescing together of diverse and perhaps vague notions of an idea into a clear and crystallized concept (an epiphany of sorts for a concept or thing).

42. R. R. Palmer, *The Age of the Democratic Revolution* (Princeton, N.J.: Princeton University Press, 1964), 54.

43. Palmer, *The Age of the Democratic Revolution,* 55.

44. Herbert Butterfield, *The Peace Tactics of Napoleon 1806–1808* (Cambridge: Cambridge University Press, 1929), 223–36, 372–80.

45. Marguerite Gerstell, "Prussian Education and Mathematics," *The American Mathematical Monthly* 82 (1975), 240–41.

46. Col. Trevor N. Dupuy, *A Genius for War: The German Army and General Staff, 1807–1945* (Englewood Cliffs, N.J.: Prentice-Hall, 1977), 22.

47. Guimps, *Pestalozzi, His Life and Work,* 257.

48. Quick, *Essays on Educational Reformers,* 347.

49. Will Seymour Monroe, *History of the Pestalozzian Movement in the United States* (Syracuse: C. W. Bardeen, 1907), 97.

50. Paul Monroe, ed. *A Cyclopedia of Education,* Vol. 4 (New York: Macmillan, 1913), 659.

51. Monroe, *A Cyclopedia of Education,* Vol. 4, 659.

52. Benjamin Duke, *The History of Modern Japanese Education: Constructing the National School System, 1872–1890* (New Brunswick, N.J.: Rutgers, 2009), 186–97.

53. John Alfred Green, *The Educational Ideas of Pestalozzi* (London: W. B. Clive, 1905), 165.

54. Kate Silber, *Pestalozzi: The Man and His Work* (New York: Schocken Books, 1973), 289.

55. Charles Edward H. Orpen, *The Pestalozzian Primer; or, the First Number of the Pestalozzian Parents' Assistant, Presenting the First Step of an Explanation of the Method of Teaching Children, in a Natural, Practical Manner (1829)* (Dublin: R. M. Tims, 1829), 15.

56. Warren Colburn, *First Lessons in Arithmetic on the Plan of Pestalozzi,* 2nd ed. (Boston: Cummings and Hilliard, 1822), xiv.

57. American Academy of Arts and Sciences, *American Academy of Arts and Sciences Academy Members 1780–2011* (Cambridge, Mass.: American Academy of Arts and Sciences, 2012), 116, 214, 350.

58. Walter S. Monroe, "Warren Colburn on the Teaching of Arithmetic Together with an Analysis of His Arithmetic Texts: I. The Life of Warren Colburn," *The Elementary School Teacher* 12 (1912): 424.

59. Theodore Edson, *Memoir of Warren Colburn: Written for the American Journal of Education* (Boston: Brown, Taggard and Chase, 1856), 13.

60. Florian Cajori, *The Teaching and History of Mathematics in the United States* (Washington, D.C.: Government Printing Office, 1890), 106.

61. Karen D. Michalowicz and Arthur C. Howard, "Pedagogy in Text: An Analysis of Mathematics Texts from the Nineteenth Century," in *A History of School Mathematics,* eds. George M. A. Stanic and Jeremy Kilpatrick (Reston, Va.: National Council of Teachers of Mathematics, 2003), 87.

62. Patricia Cline Cohen, "Numeracy in Nineteenth Century America," in *A History of School Mathematics,* eds. George M. A. Stanic and Jeremy Kilpatrick (Reston, Va.: National Council of Teachers of Mathematics, 2003), 66.

63. Andrew Wylie, "Female Influences and Education," *American Annals of Education* 8 (1838): 386.

64. Wylie, "Female Influences and Education," 387.

65. Cohen, "Numeracy in Nineteenth Century America," 69.

66. Cohen, "Numeracy in Nineteenth Century America," 60.

67. "Notices," *American Annals of Education and Instruction* 4 (1834), 148.

68. "Treatises on Algebra," *American Annals of Education* 9 (1839), 265.

69. Tayler Lewis, "Three Absurdities of Certain Modern Theories of Education," *The Biblical Repertory and Princeton Review* 23 (1851), 265–92.

70. Cohen, "Numeracy in Nineteenth Century America," 65.

71. Cohen, "Numeracy in Nineteenth Century America," 64.

72. John Dewey, *Democracy and Education: An Introduction to the Philosophy of Education* (New York: Macmillan, 1922), 11.

73. George Armitage Miller, "The Cognitive Revolution: A Historical Perspective," *Trends in Cognitive Sciences* 7 (2003): 143.

74. Liping Ma, *Knowing and Teaching Elementary Mathematics: Teachers Understanding of Fundamental Mathematics in China and the United States* (Mahwah, N.J.: Lawrence Erlbaum Associates, 1999), xvii.

75. Jacob William Albert Young, *The Teaching of Math in the Higher Schools of Prussia* (London: Longman, Greens, 1900), 106–11.

76. Quick, *Essays on Educational Reformers*, 46.

77. John McCormick, *George Santayana: A Biography* (New Brunswick, N.J.: Transaction Publishers, 2009), 144.

78. National Mathematics Advisory Panel, *Foundations for Success: The Final Report of the National Mathematics Advisory Panel* (Washington, D.C.: U.S. Department of Education, 2008).

79. Robert Goulding, *Defending Hypatia: Ramus, Savile, and the Renaissance Rediscovery of Mathematical History* (Dordrecht: Springer, 2010), 30.

80. Johann Friedrich Herbart, *The Science of Education*, Henry M. and Emmie Felkin, trans. (Boston: D. C. Heath, 1895), 40.

81. Jerome S. Bruner, *The Process of Education* (Cambridge, Mass.: Harvard University Press, 1960), 7.

82. George Abram Miller, *Historical Introduction to Mathematical Literature* (New York: Macmillan, 1916), ix–x.

83. Augustus de Morgan, *Arithmetical Books from the Invention of Printing to the Present Time: Being Brief Notices of a Large Number of Works Drawn Up from Actual Inspection* (London: Taylor and Walton, 1847), xii.

84. Douglas A. Grouws and Kristin J. Cebulla, "Elementary and Middle School Mathematics at the Crossroads," in *American Education: Yesterday, Today and Tomorrow*, ed. Thomas L. Good (Chicago: National Society for the Study of Education, 2000), 218.

85. David Eugene Smith, *The Teaching of Elementary Mathematics* (New York: Macmillan, 1906), 110.

86. Deborah Loewenberg Ball, *Summary of Testimony to U.S. House of Representatives Committee on Education and Labor* (Washington, D.C.: May 4, 2010), 1.

87. Quick, *Essays on Educational Reformers*, 191.

88. David Paul Ausubel, *Educational Psychology: A Cognitive View* (New York: Holt, Rinehart and Winston, 1968), vi.

89. Franklin Verzelius Newton Painter, *A History of Education: Revised, Enlarged and Largely Rewritten* (New York: D. Appleton, 1904), 278.

90. Carl Wieman, "Why Not Try a Scientific Approach to Science Education?" *Change—The Magazine of Higher Learning*, September–October, 2007, 9–15.

91. Israel Kleiner, *Excursions in the History of Mathematics* (New York: Springer, 2012), 268.

92. A paraphrase of the quote: "I love that expression that I heard up in Vermont originally, the lamb farmers say, You don't fatten your lambs by weighing them. But there's an awful lot of weighing of the lambs nowadays, while the lambs are getting thinner and thinner." Jonathan Kozol, "Race and Class in Public Education" (transcript from a speech delivered at State University of New York at Albany, October 17, 1997), 2. The exact words of the quote in chapter 11 are said to have been stated by Jonathan Kozol at the 157th Commencement of Westfield State College (1990s) in Westfield, Massachusetts.

93. Anna Sfard, "On Two Metaphors for Learning and the Dangers of Choosing Just One," *Educational Researcher*, March 1998, 4–13.

Part IV

1. James Thurber, *Lanterns and Lances* (New York: Time Incorporated, 1962), 121.

2. John L. Casti, *Reality Rules: Picturing the World in Mathematics—The Fundamentals*, Vol. 1 (New York: Wiley, 1997), vii.

Chapter 12

1. Raymond Louis Wilder, *Evolution of Mathematical Concepts: An Elementary Study* (New York: Wiley, 1968), 5.

2. George Abram Miller, *Historical Introduction to Mathematical Literature* (New York: Macmillan, 1916), v–vi.

3. Douglas Bernstein et al., *Psychology*, 8th ed. (Boston: Houghton Mifflin, 2008), 74.

4. The particular choice of naming is of course based on the history of this country but the conclusion is true regardless of what naming scheme is used. Adding any set of names to the map leads to a variety of questions that simply wouldn't exist without the names.

5. In this book we have called this type of counting *order counting* to distinguish it from the first method of counting we introduced in the book which we called *tally counting*.

6. Data obtained from *Airports Council International: The Voice of the World's Airports*, 2012, www.aci.aero/Data-Centre/Annual-Traffic-Data.

7. Here we have chosen to ignore the finer details of obtaining precise measurements.

8. Alfred W. Crosby, *The Measure of Reality: Quantification and Western Society, 1250–1600* (New York: Cambridge University Press, 1997), 75–93.

9. Bank information obtained from the *Federal Deposit Insurance Corporation*, www2.fdic.gov/idasp/index.asp (September 7, 2012).

10. Philip J. Davis and Reuben Hersh, *The Mathematical Experience* (Boston: Birkhauser, 1981), 399.

Chapter 13

1. Frank Smith, *Understanding Reading: A Psycholinguistic Analysis of Reading and Learning to Read* (Mahway, N.J.: Lawrence Erlbaum Associates, 2004), 192.

2. Thomas Carlyle, *Historical Essays*, ed. Chris Vanden Bossche (Berkeley: University of California Press, 2002), 4.

3. Apostolos Doxiadis, "Embedding Mathematics in the Soul: Narrative as a Force in Mathematics Education." (Text of the opening address to the Third Mediterranean Conference on Mathematics Education, Athens, Greece, January 3, 2003), 24.

4. The notion of viewing branches of mathematical knowledge as varying forms of discourses has one of its most powerful supporters in the person of researcher Anna Sfard. See Anna Sfard, *Thinking as Communicating: Human Development, the Growth of Discourses, and Mathematizing* (Cambridge: Cambridge University Press, 2008).

5. A paraphrase of Lou Ann Walker's quote: "Theories and goals of education don't matter a whit if you don't consider your students to be human beings." Lou Ann Walker, *A Loss for Words: The Story of Deafness in a Family* (New York: Harper Row, 1987), 138.

6. Alfred North Whitehead, *An Introduction to Mathematics* (New York: Henry Holt, 1911), 61.

7. John Perry, *Discussion on the Teaching of Mathematics: Which Took Place on September 14th, at a Joint Meeting of Two Sections: Section A—Mathematics and Physics; Section L—Education*, 2nd ed., edited by British Association for the Advancement of Science (London: Macmillan, 1902), 4.

Bibliography

Airports Council International: The Voice of the World's Airports, 2012, www.aci.aero/Data-Centre/ Annual-Traffic-Data.

American Academy of Arts and Sciences. *American Academy of Arts and Sciences Academy Members 1780–2011*. Cambridge (Mass.): American Academy of Arts and Sciences, 2012.

Anderson, Lewis Flint. *History of Common School Education*. New York: Henry Holt, 1909.

Atiyah, Sir Michael. "Special Article Mathematics in the 20th Century." *Bulletin of the London Mathematical Society* 34 (2002): 1–15.

Ausubel, David Paul. *Educational Psychology: A Cognitive View*. New York: Holt, Rinehart and Winston, 1968.

Ayoub, Raymond G., ed. *Musings of the Masters: An Anthology of Mathematical Reflection*. Washington, D.C.: Mathematical Association of America, 2004.

Bacon, Francis. *The New Organon*. Edited by Lisa Jardine and Michael Silverthorne. Cambridge: Cambridge University Press, 2000.

Bagley, William Chandler. *Craftsmanship in Teaching*. New York: Macmillan, 1912.

Ball, Deborah Loewenberg. "The Mathematical Understandings That Prospective Teachers Bring to Teacher Education." *The Elementary School Journal* 90 (1990): 449–66.

———. *Summary of Testimony to U.S. House of Representatives Committee on Education and Labor*. Washington, D.C.: May 4, 2010, 1–4.

Ball, Walter William Rouse. *A History of the Study of Mathematics at Cambridge*. Cambridge: Cambridge University Press, 1889.

Barnard, Francis Pierrepont. *The Casting Counter and the Counting-board: Chapter in the History of Numismatics and Early Arithmetic*. Oxford: Clarendon Press, 1916.

Bass, Hyman. "Mathematicians as Educators." In *Notices of the American Mathematical Society* 44 (1997): 18–21.

———. "Mathematics, Mathematicians and Mathematics Education." In *Bulletin (New Series) of the American Mathematical Society* 42 (2005): 417–30.

Beckmann, Petr. *A History of Pi*. New York: St. Martin's Press, 1971.

Ben-Menahem, Ari. *Historical Encyclopedia of Natural and Mathematical Sciences*. Vol. 1. Berlin: Springer-Verlag, 2009.

Bernstein, Douglas, Louis Penner, Alison Clarke-Stewart, and Edward Roy. *Psychology*. 8th ed. Boston: Houghton-Mifflin, 2008.

Bonnycastle, John. *Bonnycastle's Scholar's Guide to Arithmetic: Or, A Complete Exercise Book for the Use of Schools. With Notes Containing the Reason of Every Rule, Deduced from the Most Simple and Evident Principles.* . . . 18th ed. Edited by John Rowbotham. Philadelphia: Collins and Croft, 1818.

Boyer, Carl B. *A History of Mathematics*. 2nd ed. Revised by Uta C. Merzbach. New York: John Wiley, 1991.

Britannica Online Encyclopedia," 2012, www.britannica.com.

Bruner, Jerome S. *The Process of Education*. Cambridge, Mass.: Harvard University Press, 1960.

Bucher, Roger, Dale Griffin, and Johanna Peetz. "The Planning Fallacy: Cognitive, Motivations, and Social Origins." In *Advances in Experimental Social Psychology 43*, edited by Mark P. Zanna and James M. Olson, 1–56. Amsterdam: Elsevier, 2010.

Burton, David M. *The History of Mathematics: An Introduction*. 5th ed. New York: McGraw-Hill, 2003.

Butterfield, Herbert. *The Peace Tactics of Napoleon 1806–1808*. Cambridge: Cambridge University Press, 1929.

Cajori, Florian. "The Controversy on the Origin of Our Numerals." *The Scientific Monthly* 9 (1919): 458–64.

———. *A History of Elementary Mathematics with Hints on Methods of Teaching*. New York: Macmillan, 1929.

———. *A History of Mathematical Notations: Two Volumes Bound as One*. New York: Dover, 2007.

———. *A History of Mathematics*. 2nd ed. New York: Macmillan, 1919.

———. *The Teaching and History of Mathematics in the United States*. Washington, D.C.: Government Printing Office, 1890.

Carlyle, Thomas. *Historical Essays*. Edited by Chris Vanden Bossche. Berkeley: University of California Press, 2002.

Casti, John L. *Reality Rules: Picturing the World in Mathematics—The Fundamentals*. Vol. 1. New York: Wiley, 1997.

Chambers, Robert, ed. *Cyclopaedia of English Literature: A History, Critical and Biographical of British Authors, From the Earliest to the Present Times*. Vol. 1. Edinburgh: W. & R. Chambers, 1858.

Chesterson, Gilbert Keith. *Tremendous Trifles*. New York: Dodd, Mead, 1920.

Chrisomalis, Stephen. "The Cognitive and Cultural Foundations of Numbers." In *The Oxford Handbook of the History of Mathematics*. Edited by Eleanor Robson and Jacqueline Stedall. Oxford: Oxford University Press, 2009.

Cocker, Edward. *Cocker's Decimal Arithmetick*. Edited by John Hawkins. London: Sawbridge and Wellington, 1703.

Cohen, Patricia Cline. *A Calculating People: The Spread of Numeracy in Early America*. Chicago: University of Chicago Press, 1982.

———. "Numeracy in Nineteenth Century America." In *A History of School Mathematics*. Edited by George M. A. Stanic and Jeremy Kilpatrick, 43–71. Reston, Va.: National Council of Teachers of Mathematics, 2003.

Colburn, Warren. *First Lessons in Arithmetic on the Plan of Pestalozzi*. 2nd ed. Boston: Cummings and Hilliard, 1822.

Comenius, John Amos. *The Great Didactic of John Amos Comenius*. Translated by M. W. Keatinge. London: Adam and Charles Black, 1896.

Conant, James B. "Comenius and Harvard." In *Teacher of Nations: Addresses and Essays in Commemoration of the Visit to England of the Great Czech Educationalist Jan Amos Komenskyï, Comenius, 1641–1941*. Edited by Joseph Needham. Cambridge: Cambridge University Press, 1942.

Craig, Edward, ed. *Routledge Encyclopedia of Philosophy*. Vol. 8. London: Routledge, 1998.

Crosby, Alfred W. *The Measure of Reality: Quantification and Western Society, 1250–1600*. New York: Cambridge University Press, 1997.

Davis, Philip J., and Reuben Hersh. *The Mathematical Experience*. Boston: Birkhauser, 1981.

De Morgan, Augustus. *Arithmetical Books from the Invention of Printing to the Present Time: Being Brief Notices of a Large Number of Works Drawn Up from Actual Inspection*. London: Taylor and Walton, 1847.

Denniss, John. "Learning Arithmetic: Textbooks and Their Users in 1500–1900." In *The Oxford Handbook of the History of Mathematics*. Edited by Eleanor Robson and Jacqueline Stedall. Oxford: Oxford University Press, 2009.

Deutscher, Guy. *The Unfolding of Language: An Evolutionary Tour of Man's Greatest Invention*. New York: Henry Holt, 2005.

Devlin, Keith. *The Man of Numbers: Fibonacci's Arithmetic Revolution*. New York: Walker, 2011.

Dewey, John. *Democracy and Education: An Introduction to the Philosophy of Education*. New York: Macmillan, 1922.

Dilworth, Thomas. *The Schoolmaster's Assistant: Being a Compendium of Arithmetic Both Practical and Theoretical*. Revised by R. Tagart. New York: Daniel D. Smith, 1818.

Donnelly, Colleen. *Linguistics for Writers*. Albany: State University Press of New York, 1994.

Doxiadis, Apostolos. "Embedding Mathematics in the Soul: Narrative as a Force in Mathematics Education." Transcript of the *Opening Address to the Third Mediterranean Conference of Mathematics Education*. Athens: January 3–5, 2003.

Duke, Benjamin. *The History of Modern Japanese Education: Constructing the National School System, 1872–1890*. New Brunswick, N.J.: Rutgers University Press, 2009.

Dupuy, Col. Trevor N. *A Genius for War: The German Army and General Staff, 1807–1945*. Englewood Cliffs, N.J.: Prentice-Hall, 1977.

Eamon, William. "Markets, Piazzas, and Villages." In *The Cambridge History of Science: Early Modern Science*. Vol. 3. Edited by Katherine Park and Lorraine Daston. Cambridge: Cambridge University Press, 2006.

Edson, Theodore. *Memoir of Warren Colburn: Written for the American Journal of Education*. Boston: Brown, Taggard and Chase, 1856.

Edwards, Tryon. *A Dictionary of Thoughts: Being a Cyclopedia of Laconic Quotations from the Best Authors of the World, Both Ancient and Modern*. Detroit: F. B. Dickerson, 1908.

Einstein, Albert. *Cosmic Religion with Other Opinions and Aphorisms*. New York: Covici-Friede, 1931.

Ellerton, Nerida, and Ken Clements. "The Process of Decolonizing School Mathematics Textbooks and Curricula in the United States." *ICME* 11, TSG 38 (2008), 1–5. http://tsg.icme11.org/document/get/783.

Estrada, Ernesto. *The Structure of Complex Networks: Theory and Applications*. New York: Oxford University Press, 2011.

Farrington, Benjamin. *Greek Science: Its Meaning for Us*. Baltimore: Penguin Books, 1969.

Fauvel, John, and Jan Van Maanen, eds. *History in Mathematics Education: The ICMI Study*. Dordrecht: Kluwer Academic Publishers, 2000.

Febvre, Lucien, and Henri-Jean Martin. *The Coming of the Book: The Impact of Printing 1450–1800*. Translated by David Gerard. London: Verso, 1997.

Feynman, Richard. *Six Easy Pieces*. New York: Basic Books, 2011.

Fibonacci, Leonardo, and Laurence Sigler. *Fibonacci's Liber Abaci: A Translation into Modern English of Leonardo Pisano's Book of Calculation.* New York: Springer, 2002.

Fink, Karl. *A Brief History of Mathematics (Geschichte der Elementar-Mathematik).* 2nd ed. Translated by Wooster Woodruff Beman and David Eugene Smith. Chicago: Open Court, 1903.

Fitch, Joshua G. *Lectures on Teaching: Delivered in the University of Cambridge during the Lent Term, 1880.* New York: Macmillan, 1891.

Fitch, Sir Joshua G. *Educational Aims and Methods: Lectures and Addresses.* New York: Macmillan, 1900.

Flanagan, Frank M. *The Greatest Educators Ever.* London: Continuum, 2006.

Franklin, Benjamin. *Autobiography of Benjamin Franklin.* Edited by Charles W. Eliot. New York: P. F. Collier, 1909.

Freudenthal, Hans. *Mathematics as an Educational Task.* Dordrecht: D. Reidel Publishing, 1973.

Fromkin, Victoria, Robert Rodman, and Nina Hyams. *An Introduction to Language.* 9th ed. Boston: Wadsworth, 2011.

Gaither, Carl C., and Alma E. Cavazos-Gaither, eds. *Gaither's Dictionary of Scientific Quotations Second Edition.* New York: Springer, 2012.

Gelb, Ignace J. *A Study of Writing.* Chicago: University of Chicago Press, 1952.

Gerstell, Marguerite. "Prussian Education and Mathematics." *The American Mathematical Monthly* 82 (1975): 240–45.

Glenn, Jason. *Politics and History in the Tenth Century: The Work and World of Richer of Riems.* New York: Cambridge University Press, 2004.

Gnanadesikan, Amalia E. *The Writing Revolution: Cuneiform to the Internet.* Oxford: Blackwell-Wiley, 2008.

Goodrich, Samuel Griswold. *The Child's Arithmetic: Being an Easy and Cheap Introduction to Daboll's, Pike's, White's and other Arithmetics.* Hartford: Samuel Griswold Goodrich, 1818.

Goulding, Robert. *Defending Hypatia: Ramus, Savile, and the Renaissance Rediscovery of Mathematical History.* Dordrecht: Springer, 2010.

Grant, Hardy. "A Sketch of the Cultural Career of Mathematics." In *Essays in Humanistic Mathematics.* Edited by Alvin White, 21–28. Washington, D.C.: Mathematical Association of America, 1993.

Green, John Alfred. *The Educational Ideas of Pestalozzi.* London: W. B. Clive, 1905.

Grendler, Paul F. *Books and Schools in the Italian Renaissance.* Rugby, UK: Variorum, 1995.

———. *Schooling in Renaissance Italy, Literacy and Learning 1300–1600.* Baltimore: Johns Hopkins University Press, 1991.

———. *The Universities of the Italian Renaissance.* Baltimore: Johns Hopkins University Press, 2002.

Grouws, Douglas A., and Kristin J. Cebulla. "Elementary and Middle School Mathematics at the Crossroads." In *American Education: Yesterday, Today and Tomorrow.* Edited by Thomas L. Good, 209–55. Chicago: National Society for the Study of Education, 2000.

Guimps, Roger de. *Pestalozzi: His Life and Work.* 2nd ed. Translated by J. Russell. New York: D. Appleton, 1890.

Hadden, Richard W. *On the Shoulders of Merchants: Exchange and the Mathematical Conception of Nature in Early Modern Europe.* Albany: State University Press of New York, 1994.

Hardy, G. H. *A Mathematician's Apology.* Cambridge: Cambridge University Press, 1999.

Harris, Roy. *Signs of Writing.* New York: Routledge, 1995.

———. *Signs, Language and Communication.* New York: Routledge, 1996.

———. *Rethinking Writing.* London: Continuum, 2001.

Heeffer, Albrecht. "The Abbacus Tradition: The Missing Link Between Arabic and Early Symbolic Algebra?" Presented at International Seminar on the History of Mathematics in Memory of Subhash Handa. New Delhi: December 17–18, 2007.

Herbart, Johann Friedrich. *The Science of Education.* Translated by Henry M. and Emmie Felkin. Boston: D. C. Heath, 1895.

Hogben, Lancelot. *Mathematics for the Million.* New York: W. W. Norton, 1993.

Holman, Henry. *Pestalozzi: An Account of His Life and Work.* London: Longmans, Green, 1908.

Howard, Ron. *Apollo 13.* Universal Pictures, 1995.

Howson, Geoffrey. *A History of Mathematics Education in England.* Cambridge: Cambridge University Press, 1982.

Hutton, Charles. *The Schoolmasters Guide, or a Complete System of Practical Arithmetic: Adapted to the Use of Schools.* 2nd ed. Newcastle upon Tyne: J. White and T. Saint for R. Baldwin, 1766.

Ifrah, Georges. *The Universal History of Numbers: From Prehistory to the Invention of the Computer.* New York: Wiley, 2000.

Issawi, Charles. "Europe, the Middle East and the Shift in Power: Reflections on a Theme by Marshall Hodgson." *Comparative Studies in Society and History 22* (1980): 487–504.

Jackson, Lambert Lincoln. *The Educational Significance of 16th Century Arithmetic: From the Point of View of the Present Time.* New York: Teacher's College, Columbia University, 1906.

Jones, Henry. "The Social Organism." In *The British Idealists.* Edited by David Boucher. New York: Cambridge University Press, 1997.

Karmiloff, Kyra, and Annette Karmiloff-Smith. *Pathways to Language: From Fetus to Adolescent.* Cambridge, Mass.: Harvard University Press, 2001.

Kidwell, Peggy Aldrich, Amy Ackerberg-Hastings, and David Lindsay Roberts. *Tools of American Mathematics Teaching: 1800–2000.* Washington, D.C.: Johns Hopkins University Press and Smithsonian Institution, 2008.

Klein, Jacob. *Greek Mathematical Thought and the Origin of Algebra.* Cambridge, Mass.: MIT Press, 1968.

Kleiner, Israel. *Excursions in the History of Mathematics.* New York: Springer, 2012.

Kojima, Takashi. *The Japanese Abacus: Its Use and Theory.* Tokyo: Tuttle, 1954.

Kozol, Jonathan. "Race and Class in Public Education." Transcript from a speech delivered at State University of New York at Albany, October 17, 1997.

Krusi, Hermann, Jr. *Pestalozzi.* Carlisle, Mass.: Applewood Books, 1875.

Lakoff, George, and Rafael Nunez. *Where Mathematics Comes From.* New York: Basic Books, 2000.

Lebesgue, Henri. *Measure and the Integral.* Edited by Kenneth O. May. San Francisco: Holden-Day, 1966.

Leinwand, Steven. "It's Time to Abandon Computational Algorithms." *Education Week*, 13 (1994): 36.

Lemos, Marion de. *Closing the Gap between Research and Practice: Foundations for the Acquisition of Literacy.* Camberwell: Australian Council for Educational Research, 2002.

Lewis, Tayler. "Three Absurdities of Certain Modern Theories of Education." *The Biblical Repertory and Princeton Review* 23 (1851): 265–92.

Littlefield, George Emery. *Early Schools and School-Books of New England.* Boston: Club of Odd Volumes, 1904.

Ma, Liping. *Knowing and Teaching Elementary Mathematics: Teachers Understanding of Fundamental Mathematics in China and the United States.* Mahwah, N.J.: Lawrence Erlbaum Associates, 1999.

Manin, Yuri I. *Mathematics as Metaphor.* Providence, R.I.: American Mathematical Society, 2007.

Mann, Horace, ed. *The Common School Journal* 5 (1843): 119.

"Math Made Interesting." *Time*, September 22, 1961.

Maynes, Mary Jo. *Schooling for the People: Comparative Local Studies of Schooling History in France and Germany 1750–1850.* New York: Holmes and Meier, 1985.

McCormick, John. *George Santayana: A Biography.* New Brunswick, N.J.: Transaction Publishers, 2009.

Menninger, Karl. *Number Words and Number Symbols: A Cultural History of Numbers.* Cambridge, Mass.: MIT Press, 1969.

"Merriam-Webster Online Dictionary." 2012, www.merriam-webster.com.

Michalowicz, Karen D., and Arthur C. Howard. "Pedagogy in Text: An Analysis of Mathematics Texts from the Nineteenth Century." In *A History of School Mathematics.* Edited by George M. A. Stanic and Jeremy Kilpatrick, 77–106. Reston, Va.: National Council of Teachers of Mathematics, 2003.

Miller, George Abram. *Historical Introduction to Mathematical Literature.* New York: Macmillan, 1916.

Miller, George Armitage. "The Cognitive Revolution: A Historical Perspective." *Trends in Cognitive Sciences* 7 (2003): 141–44.

———. *The Science of Words.* New York: Scientific American Library, 1991.

Monroe, Paul, ed. *A Cyclopedia of Education.* Vol. 4. New York: Macmillan, 1913.

———, ed. *A Cyclopedia of Education.* Vol. 5. New York: Macmillan, 1913.

Monroe, Walter S. "Warren Colburn on the Teaching of Arithmetic Together with an Analysis of His Arithmetic Texts: I. The Life of Warren Colburn." In *The Elementary School Teacher* 12 (1912), 421–26.

———. "Warren Colburn on the Teaching of Arithmetic Together with an Analysis of His Arithmetic Texts: II. Teaching of Arithmetic." In *The Elementary School Teacher* 12 (1912), 463–80.

Monroe, Will Seymour. *Comenius and the Beginnings of Educational Reform.* New York: Charles Scribner, 1900.

———. *History of the Pestalozzian Movement in the United States.* Syracuse: C. W. Bardeen, 1907.

Morgan, Thomas J. *Educational Mosaics: A Collection from Many Writers (Chiefly Modern) of Thoughts Bearing on Educational Questions of the Day.* Boston: Silver, Rogers, 1887.

Moynahan, Brian. *God's Bestseller: William Tyndale, Thomas More, and the Writing of the English Bible—A Story of Martyrdom and Betrayal.* New York: St. Martin's Press, 2003.

Munoz, Peter. *Frederick Barbarossa: A Study in Medieval Politics.* Ithaca, N.Y.: Cornell University Press, 1969.

National Mathematics Advisory Panel. *Foundations for Success: The Final Report of the National Mathematics Advisory Panel.* Washington, D.C.: U.S. Department of Education, 2008.

Neal, Katherine. *From Discrete to Continuous: The Broadening of Number Concepts in Early Modern England.* Dordrecht: Kluwer Academic Publishers, 2002.

Neugebauer, Otto. *Exact Sciences in Antiquity.* 2nd ed. New York: Dover, 1969.

Ness, Harald. "Mathematics: An Integral Part of Our Culture." In *Essays in Humanistic Mathematics,* 49–52. Edited by Alvin White. Washington, D.C.: Mathematical Association of America, 1993.

Nietzsche, Friedrich Wilhelm. *Twilight of the Idols: With the Antichrist and Ecce Homo.* Translated by Antony M. Ludovici. Ware, UK: Wordsworth Editions Limited, 2007.

O'Connor, J. J., and E. F. Robertson. "Brahmagupta," on *MacTutor History of Mathematics,* www-history.mcs.st-andrews.ac.uk/Biographies/Brahmagupta.html, 2000.

O'Connor, J. J., and E. F. Robertson. "Arabic Mathematics: Forgotten Brilliance?" on *MacTutor History of Mathematics*, www-history.mcs.st-and.ac.uk/HistTopics/Arabic_mathematics. html, 1999.

Olson, David R. *The World on Paper: The Conceptual and Cognitive Implications of Writing and Reading*. New York: Cambridge University Press, 1996.

Orpen, Charles Edward H. *The Pestalozzian Primer; or, the First Number of the Pestalozzian Parents' Assistant, Presenting the First Step of an Explanation of the Method of Teaching Children, in a Natural, Practical Manner (1829)*. Dublin: R. M. Tims, 1829.

Painter, Franklin Verzelius Newton. *A History of Education: Revised, Enlarged and Largely Rewritten*. New York: D. Appleton, 1904.

Palmer, R. R. *The Age of the Democratic Revolution*. Princeton, N.J.: Princeton University Press, 1964.

Perera, Katharine. *Reading and Writing*. Oxford: Blackwell, 1989.

Perry, John. *Discussion on the Teaching of Mathematics: Which Took Place on September 14th, at a Joint Meeting of Two Sections: Section A—Mathematics and Physics; Section L—Education*. 2nd ed. Edited by British Association for the Advancement of Science. London: Macmillan, 1902.

Pestalozzi, Johann Heinrich. *How Gertrude Teaches Her Children: An Attempt to Help Mothers to Teach Their Own Children and an Account of the Method—A Report to the Society of the Friends of Education, Burgdorf*. 2nd ed. Translated by Lucy E. Holland and Frances C. Turner. Edited by Ebenezer Cook. Syracuse: C. W. Bardeen, 1898.

———. *Leonard and Gertrude*. Translated and abridged by Eva Channing. Boston: D.C. Heath, 1906.

Petty, Sir William. "Plan of a Trade or Industrial School." *The American Journal of Education* XI (1862), 199–208.

Pike, Nicholas. *A New and Complete System of Arithmetick: Composed for the Use of the Citizens of the United States*. 8th ed. Revised by Nathaniel Lord. New York: Evert Duyckinck, 1816.

Pillis, John de. *777 Mathematical Conversation Starters*. Washington, D.C.: Mathematical Association of America, 2002.

Pinker, Steven. *The Language Instinct: How the Mind Creates Language*. New York: Perennial Classics, 2000.

———. *The Stuff of Thought: Language as a Window into Human Nature*. New York: Viking, 2007.

Plofker, Kim. *Mathematics in India*. Princeton, N.J.: Princeton University Press, 2008.

Pollack, Henry O. "A History of Teaching Modeling." In *A History of School Mathematics*. Edited by George M. A. Stanic and Jeremy Kilpatrick, 647–69. Reston, Va.: National Council of Teachers of Mathematics, 2003.

Powell, Barry B. *Writing: Theory and History of the Technology of Civilization*. Chichester, UK: Wiley-Blackwell, 2009.

Quick, Robert Hebert. *Essays on Educational Reformers*. New York: D. Appleton, 1897.

Reiss, Timothy J. *Knowledge, Discovery and Imagination in Early Modern Europe: The Rise of Aesthetic Rationalism*. Cambridge: Cambridge University Press, 1997.

Ross, Susan, and Mary Pratt-Cotter. "Subtraction in the United States: An Historical Perspective." *The Mathematics Educator* 8 (1997): 4–8.

Sampson, Geoffrey. *Writing Systems: A Linguistic Introduction*. Stanford, Calif.: Stanford University Press, 1985.

Sanford, Vera. *A Short History of Mathematics*. Edited by John Wesley Young. Boston: Houghton Mifflin, 1930.

Schmandt-Besserat, Denise. *How Writing Came About*. Austin: University of Texas Press, 1996.

Sfard, Anna. "On Two Metaphors for Learning and the Dangers of Choosing Just One." *Educational Researcher*, March 1998, 4–13.

———. *Thinking as Communicating: Human Development, the Growth of Discourses, and Mathematizing.* Cambridge: Cambridge University Press, 2008.

Sihler, Andrew L. *Language History: An Introduction.* Amsterdam: John Benjamins Publishing, 2000.

Silber, Kate. *Pestalozzi: The Man and His Work.* New York: Schocken Books, 1973.

Smith, David Eugene. *The History of Mathematics.* Vol. 2. Edited by Eva May Luse Smith. New York: Dover, 1958.

———. *A Source Book for Mathematics: 125 Classic Selections from the Writings of Pascal, Leibniz, Euler, Fermat, Gauss, Descartes, Newton, Riemann and Many Others.* New York: Dover, 1959.

———. *The Teaching of Elementary Mathematics.* New York: Macmillan, 1906.

Smith, David Eugene, and Jekuthiel Ginsburg. "*From Numbers to Numerals and From Numerals to Computation.*" In *The World of Mathematics.* Vol. 1. Edited by James R Newman, 442–64. New York: Simon & Schuster, 1956.

Smith, David Eugene, and Louis Charles Karpinski. *The Hindu Arabic Numerals.* Boston: Ginn, 1911.

Smith, Frank. *Understanding Reading: A Psycholinguistic Analysis of Reading and Learning to Read.* Mahway, N.J.: Lawrence Erlbaum Associates, 2004.

Steele, Robert. *Earliest Arithmetics in English.* London: Oxford University Press, 1922.

Steiner, Christopher B. "Authenticity, Repetition, and the Aesthetics of Seriality: The Work of Tourist Art in the Age of Mechanical Reproduction." In *Unpacking Culture: Art and Commodity in Colonial and Postcolonial Worlds.* Edited by Ruth B. Phillips and Christopher B. Steiner, 100–101. Berkeley: University of California Press, 1999.

Stewart, Ian. *From Here to Infinity.* New York: Oxford University Press, 1996.

Stoner, Gregory N., and Patricia McCarthy. "The Market for Luca Pacioli's Summa Arithmetica." In *Accounting Historians Journal* 35 (2008), 111–34.

Swetz, Frank J. *Capitalism and Arithmetic: The New Math of the 15th Century.* LaSalle, Ill.: Open Court, 1987.

Swetz, Frank, and Victor Katz. "Opus Arithmetica of Honoratus." In *Mathematical Treasures.* Mathematical Sciences Digital Library. Washington, DC: Mathematical Association of America, 2012.

Tall, David. "The Transition to Advanced Mathematical Thinking: Functions, Limits, Infinity and Proof." In *Handbook of Research on Mathematics Teaching and Learning.* Edited by D. A. Grouws, 495–511. New York: Macmillan, 1992.

Tanton, James Stuart. *Encyclopedia of Mathematics.* New York: Facts on File, 2005.

"Treatises on Algebra." *American Annals of Education* 9 (1839): 263–68.

Vives, Juan Luis, and Watson, Foster. *Vives on Education: A Translation of the De Tradendis Disciplinis of Juan Luis Vives.* London: Cambridge University Press, 1913.

Wakeley, Ann, Susan Rivera, and Jonas Langer. "Not Proved: Reply to Wynn." *Child Development* 71 (2000): 1537–39.

Walker, Lou Ann. *A Loss for Words: The Story of Deafness in a Family.* New York: Harper Row, 1987.

Ward, James. "Notes on Education Values." *Journal of Education, A Monthly Record and Review* 22 (January–December 1890): 627–32.

Whitehead, Alfred North. *An Introduction to Mathematics.* New York: Henry Holt, 1911.

Wieman, Carl. "Why Not Try a Scientific Approach to Science Education?" *Change—The Magazine of Higher Learning* (September–October 2007): 9–15.

Wilder, Raymond Louis. *Evolution of Mathematical Concepts: An Elementary Study.* New York: Wiley, 1968.

Wilford, John Noble. "Who Began Writing? Many Theories, Few Answers." *New York Times,* April 6, 1999.

Willemsen, AnneMarieke. *Back to the Schoolyard: The Daily Practice of Medieval and Renaissance Education.* Turnhout, Belgium: Brepolis, 2008.

Wilson, Nigel, ed. *Encyclopedia of Ancient Greece.* New York: Routledge, 2006.

Witt, Ronald G. *In the Footsteps of the Ancients: The Origins of Humanism from Lovato to Bruni.* Leiden: Brill, 2000.

Wright, John W., ed. *The New York Times Guide to Essential Knowledge.* New York: St. Martin's Press, 2011.

Wylie, Andrew. "Female Influences and Education." *American Annals of Education* 8 (1838): 386.

Wynn, Karen. "Addition and Subtraction by Human Infants." *Nature* 358 (1992): 749–50.

Yong, Lam Lay. "Computation: Chinese Counting Rods." In *Encyclopedia of the History of Science, Technology, and Medicine in Non-Western Cultures,* edited by Helaine Selin, 233–34. Dordrecht: Kluwer Academic Publishers, 1997.

———. "The Development of Hindu-Arabic and Traditional Chinese Arithmetic." *Chinese Science* 13 (1996): 45.

———. "On the Chinese Origin of the Galley Method of Arithmetical Division." *British Journal of the History of Science* 3 (1966): 66–69.

Yong, Lam Lay, and Ang Tian Se. *Fleeting Footsteps: Tracing the Conception of Arithmetic and Algebra in Ancient China.* Rev. ed. Singapore: World Scientific Publishing, 2004.

Young, Jacob William Albert. *The Teaching of Math in the Higher Schools of Prussia.* London: Longman, Greens, 1900.

Yule, George. *The Study of Language.* 2nd ed. New York: Cambridge University Press, 1996.

Index

abacus, 189, 287n6; advantages of, 46–47; ancient Roman, 44, 291n7; Chinese or Suan Pan, 44–45, *44–45*, 291n7; contests, 195–96, *197*, 292n32; counter, 196, *198–99*, 200, *202*, 205, 209, 210, 211, 291n7; disadvantages of, 46–47, 200; explanation of, 41–43; HA numerals and, 51–54, 102; history, 44–45, 288n2; line, 200; memorization needed with, 47, 51, 53, 80, 200; script of, 47–54; speed, electronic calculator and, 195–96, *197*, 292n32

abacus rods: addition table, *81*, 82; addition with, 78–83, *81, 83, 84*, 259; multiplication table, *115–16*, 117; with place-value multiplication, 110–14, *115–16*, 117

abbaco schools, 205, 207, 210

abbreviations: with coin numeral system, 27; multiplication, 101–2

abstractions: arithmetic as, 225; teaching about, 244

accounting, societal, 22

Adams, Douglas, 219

a danda (method which gives), 188

addition: abacus rod addition table, *81*, *82*; with abacus rods, 78–83, *81, 83, 84*, 259; carrying in, 39–40, 43, 80, 82, 134; with coin numeral system, 36–38, 39, 78, 259; commutative property of, 103, 288n3; distributive property of multiplication over, 106–9, 113; expanded *versus* compact, 129; HA addition table, *82*, 83, 85, 86; with HA numerals, 78–80, 82–83, *82–84*, 85–86; with language number words, 36, 38–39; multiplication solved with, 99–102; with numerical coefficients, 177; with system of physical objects, *40*, 40–41

additive numeral systems, 33

advancements, in civilizations, 269

advantages: of abacus, 46–47; of using visible marks with communication, 22–23

age, number names and, 253

Ahmes (Egyptian scribe), 106

Alexandria, 64, 68, 69. *See also* Eratosthenes of Alexandria

algebra, Arabs and, 201

About the Author

G. ARNELL WILLIAMS (MS Yale) is an associate professor of mathematics at San Juan College in New Mexico. He has won numerous teaching awards for helping people to overcome their fear of math. He is also an avid landscape photographer; his images may be seen at www.dramainnature.com.